AIRCRAFT AND AVIONICS RESEARCH AND TECHNOLOGY

AIRCRAFT ELECTRONIC WARFARE

A MEMOIR BY A PROJECT ENGINEER

AIRCRAFT AND AVIONICS RESEARCH AND TECHNOLOGY

Additional books and e-books in this series can be found on Nova's website under the Series tab.

AIRCRAFT AND AVIONICS RESEARCH AND TECHNOLOGY

AIRCRAFT ELECTRONIC WARFARE

A MEMOIR BY A PROJECT ENGINEER

JOHN EDWARD WATERS

Copyright © 2019 by Nova Science Publishers, Inc.

All rights reserved. No part of this book may be reproduced, stored in a retrieval system or transmitted in any form or by any means: electronic, electrostatic, magnetic, tape, mechanical photocopying, recording or otherwise without the written permission of the Publisher.

We have partnered with Copyright Clearance Center to make it easy for you to obtain permissions to reuse content from this publication. Simply navigate to this publication's page on Nova's website and locate the "Get Permission" button below the title description. This button is linked directly to the title's permission page on copyright.com. Alternatively, you can visit copyright.com and search by title, ISBN, or ISSN.

For further questions about using the service on copyright.com, please contact:
Copyright Clearance Center
Phone: +1-(978) 750-8400   Fax: +1-(978) 750-4470   E-mail: info@copyright.com.

NOTICE TO THE READER

The Publisher has taken reasonable care in the preparation of this book, but makes no expressed or implied warranty of any kind and assumes no responsibility for any errors or omissions. No liability is assumed for incidental or consequential damages in connection with or arising out of information contained in this book. The Publisher shall not be liable for any special, consequential, or exemplary damages resulting, in whole or in part, from the readers' use of, or reliance upon, this material. Any parts of this book based on government reports are so indicated and copyright is claimed for those parts to the extent applicable to compilations of such works.

Independent verification should be sought for any data, advice or recommendations contained in this book. In addition, no responsibility is assumed by the Publisher for any injury and/or damage to persons or property arising from any methods, products, instructions, ideas or otherwise contained in this publication.

This publication is designed to provide accurate and authoritative information with regard to the subject matter covered herein. It is sold with the clear understanding that the Publisher is not engaged in rendering legal or any other professional services. If legal or any other expert assistance is required, the services of a competent person should be sought. FROM A DECLARATION OF PARTICIPANTS JOINTLY ADOPTED BY A COMMITTEE OF THE AMERICAN BAR ASSOCIATION AND A COMMITTEE OF PUBLISHERS.

Additional color graphics may be available in the e-book version of this book.

Library of Congress Cataloging-in-Publication Data

ISBN: 978-1-53616-101-4
Library of Congress Control Number:2019948409

Published by Nova Science Publishers, Inc. † New York

CONTENTS

Preface		xi
Synopsis		xiii
Chapter 1	Association of Old Crows	1
Chapter 2	Chaff in World War II	3
Chapter 3	D-Day Painting	5
Chapter 4	SGT Leonard Lomell Hero	7
Chapter 5	Pointe du Hoc	9
Chapter 6	Sainte-Mere-Eglise	11
Chapter 7	White House Invitations	13
Chapter 8	U-2 Aircraft	17
Chapter 9	High- and Low-Altitude Aircraft	19
Chapter 10	Missile Alert Special Test	21
Chapter 11	Fan and Special Test Equipment	23
Chapter 12	Canada Briefing	25
Chapter 13	First Integrated Circuit at Fort Monmouth, N.J.	27
Chapter 14	Radar Signal Detector Set AN/APR-39	29

Contents

Chapter 15	Two Radar Signals Detected and Countered	**33**
Chapter 16	Control Panel and Indicator of AN/APR-39	**35**
Chapter 17	Woman and AN/APR-39 Photo	**37**
Chapter 18	Fast Patrol Boat	**39**
Chapter 19	AN/APR-39 on Navy Seals FPB	**41**
Chapter 20	American Embassy in Iran, Missions 1 and 2	**43**
Chapter 21	Block Diagram of the AN/APR-39 (V1)	**45**
Chapter 22	V1 and V2 of the AN/APR-39	**47**
Chapter 23	OH-6 Helicopter Flight Test and Vietnam	**49**
Chapter 24	Small Pencil for Pre-Flight Check Out	**51**
Chapter 25	Chaff and Flare Dispenser on Aircraft	**53**
Chapter 26	Dual Dispenser on an Aircraft	**55**
Chapter 27	Russian Gun Dish Radar System	**57**
Chapter 28	Current Gun Dish	**59**
Chapter 29	Soviet SA-7 Heat-Seeker Missile in Vietnam	**61**
Chapter 30	Soviet SA-2 Missile System in Vietnam	**63**
Chapter 31	Cruise Missile Jammer for Space and Missile Systems Center and Strategic Defense Initiative	**65**
Chapter 32	Current Russian Jammer	**69**
Chapter 33	Flight Line Test at Lakehurt NAS, N.J.	**71**
Chapter 34	Countermeasures Set AN/ALQ-80	**73**
Chapter 35	In-Board Jammer	**77**
Chapter 36	Velocity-Timed Fuse Jammer	**79**
Chapter 37	Lt. Dennis W. Zilinski II	**81**

Contents vii

Chapter 38	Top Secret Meeting at the Sensitive Compartment Information Facility	83
Chapter 39	Firefinder AN/TPQ-36 in Afghanistan	85
Chapter 40	Firefinder AN/TPQ-37	87
Chapter 41	Old Crow Figurine	89
Chapter 42	Cold-Weather Test Alaska	91
Chapter 43	Slingshot	93
Chapter 44	Undercover Agent for a Government Agency	95
Chapter 45	Vietnam and Thailand	99
Chapter 46	Joint Special Operations Command and UAVs and Satellite Transceivers	103
Chapter 47	Germany and Germans	107
Chapter 48	No Spare Parts or Aircraft Fuel	109
Chapter 49	Fort Rucker Complaint	113
Chapter 50	Navigation Aid for Emergencies	115
Chapter 51	Italian Fascism	117
Chapter 52	Vietnam Urgency	119
Chapter 53	New Radar System	123
Chapter 54	Velocity Timed Fuse Live Test	127
Chapter 55	Honduras Briefing	129
Chapter 56	Panama Canal Zone Briefing	131
Chapter 57	Night Vision Goggles	133
Chapter 58	Pearl Harbor	137
Chapter 59	Jay Waters' Retirement	139
Chapter 60	My Retirement Plaque	141

viii *Contents*

Chapter 61	Post Retirement	**143**
Addendum 1 through 39		**145**
About the Author		**197**
Index		**199**
Related Nova Publications		**205**

PREFACE

There are many books written by editors about Electronic Warfare. This book is written by an author about his Electronic Warfare systems projects, tasks and stories.

He was a project engineer for Electronic Warfare systems for more than twenty five years and a consulting engineer for other systems for five years.

The book contains 61 chapters with 59 illustrations. There are chapters that report his special projects with the Joint Special Operations.

The author's special projects with Joint Special Operations Command, the Space & Missile Systems Center, NASA, the Navy SEALS, the CIA, the Army Delta Force, and the Army Missile Command are detailed along with his actions as an undercover agent for a national crime fighting agency. He explains why he expedited a contract award for 2,000 radar locator systems, (AN/APR-39), then had them installed in Army helicopters before the possible start of World War III.

This book describes the author's experiences being invited to visit the White House followed by a productive outcome as a result of that meeting; a second invitation when his daughter was a White House volunteer; and his trip to Thailand to evaluate the USAF latest radar warning receiver for possible use on Army fixed wing aircraft. He explains why he was not allowed to deplane after landing in Vietnam and his proposal for a barrier on the Mexican border.

A few stories about his son Colonel John (Jay) Waters are included. One touching upon Jay's story about when he was on the general's staff in Afghanistan and the unique action taken by their general that stopped the mortar firings into Bagram Army base.

A story about his friend Leonard Lomell, who a historian called the person most responsible for the success in the D-Day invasion, second only to General Eisenhower, is detailed. A ten-day visit to the Normandy beaches and cemeteries with Jay when he was the military head of Arlington National Cemetery and all world-wide Army cemeteries is also mentioned.

The author also dives into Nap-of-the-Earth flight tests while flying over the Chesapeake Bay, his pilot being a POW in North Vietnam and what that man said about Jane Fonda.

A flight test across the Fulda Gap in Germany to detect enemy radars is described along with what his pilot told him about a young couple who tried to cross the border. Then a flight test along the DMZ in North Korea and a special test in England against a new radar is mentioned.

A report about a helicopter unable to return to the airport due to a dense fog is included and how his on-board radar locator assisted the pilot to return to the airport.

When Ayatollah Khomeini became the ruler of Iran, students attacked the American Embassy. They held many Americans prisoners. The author and his comrades attempted a rescue mission, but two of their aircrafts collided in the Iranian desert and the mission was aborted. The world knew about that failed mission, but the world never knew about the second United States rescue mission and how he was a technical advisor for said second rescue mission.

SYNOPSIS

After his Navy enlistment and college, the author's career was as an Army Civilian Project Engineer for Electronic Warfare systems for more than twenty years. Then five years as a consulting engineer for other military systems.

The book contains 61 brief original chapters with 59 images for easy reading and viewing. He has chapters that report his special projects with the Joint Special Operations Command, the Space & Missile Systems Center, NASA, the Navy SEALS, CIA, Army Delta Force, and the Army Missile Command.

He explains why he expedited a contract award for 2,000 radar locator systems, (AN/APR-39), then had them installed in Army helicopters before the possible start of World War III.

He describes an invitation to visit the White House that produced a positive outcome as a result of that meeting and a second invitation when my daughter was a White House volunteer.

Stories from a POW pilot as we flew low altitude test flights above the Chesapeake Bay and why we broke our pencils in half prior to our first flight are discussed.

The author details the use of his radar locator to return a helicopter to the airport after the pilot could not see his ground visual aids due to a dense fog.

The unsuccessful mission to free the hostages in Iran that was aborted when two aircraft collided in the Iranian desert is described. The world knew about that failed mission but did not know about the second mission to free the hostages. The author was a technical advisor for the second mission.

The book also includes an amusing meeting on a YMCA pool deck with the world record holder for the breast stroke followed by a meeting with a US Military Academy, West Point cadet that had a sad ending.

The author recalls teaching his radar locator operations at worldwide briefings to pilots and his added unofficial tactics recommendations.

The author describes his friend Leonard Lomell, who a historian called the person most responsible for our successful landings on D-Day, second only to General Eisenhower, and what he did to earn that accolade.

Countermeasures against the Soviet SA-7 heat seeking missile in Vietnam are mentioned.

The author touches upon his son, Colonel Jay's, deployment to Bagram Air Base, Afghanistan and the base commander's unique solution that stopped the mortar shelling.

The author's recommendation to the Army Communications and Electronics Command for a contract award for a specific Remotely Piloted Vehicle is detailed.

His actions as an undercover agent for a national counter criminal agency are described along with a Cold Weather Test and a meeting in Alaska, meetings in Honduras and the Panama Canal Zone, a TOP SECRET meeting in the Sensitive Compartment Information Facility and the author's proposal for a barrier on the Mexican border.

Chapter 1

ASSOCIATION OF OLD CROWS

The Association of Old Crows is an International Electronic Warfare organization. I was a president of the Garden State Chapter. Camp Evans in Wall Township, N. J., has the InfoAge Science Center with the AOC/EW museum that contains literature and artifacts about Electronic Warfare.

The AOC had annual conventions in Washington, D.C. Once the convention was in Dallas so I decided to bring my wife, Barbara. A jogging friend recommended a babysitter for our three young children. When Mrs. Kay arrived, I was concerned. She was small, thin and in her seventies. But I soon knew that she would be a good babysitter.

In Dallas we stayed in an upscale downtown hotel. When we left the hotel and returned, someone had removed Barbara's money from her wallet that she left in the room. Ever since we always turn on the TV when we leave a hotel so it will sound like someone is in the room and deter intruders.

Barbara and the other spouses rode in a charter bus to the Dallas football stadium for a tour and lunch. In the evening we were transported to a ranch for a BBQ and "mountain oysters." After a couple of days Barbara did not think about the children so we visited New Orleans for two days.

Figure 1. Association of Old Crows /Garden State Chapter.

Chapter 2

CHAFF IN WORLD WAR II

In World War II the British Electronic Warfare personnel were called the Ravens. Winston Churchill called Electronic Warfare the Wizards War.

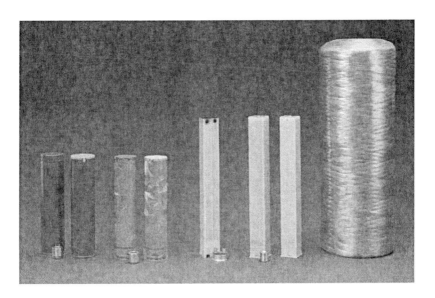

Figure 2. Chaff cartridges, squibs and a spool of chaff.

When our bombers flew missions into Germany, they dropped chaff during the mission. The only aircraft that dropped the chaff was far in front of the formation and the prime target for enemy ground fire and fighter aircraft. The uncle of our chief engineer, Ray Irwin, was a pilot for those missions. Ray said that they drew straws to decide who had to fly that lead aircraft. They knew that the aircraft was going to be hit and their best hope for the crew was to be POWs until the end of the war.

The length of each chaff strand was cut to the wavelength of the radar used by the anti-aircraft artillery operators. In the illustration, there are different length chaff cartridges and a spool of the chaff. Note the explosive squibs in front of the cartridges.

Chapter 3

D-Day Painting

Point du Hoc is on the beach in Normandy. It was one of the most bombed areas during World War II. Hours before the D-Day invasion began, a company of Rangers climbed Point du Hoc. In the drawing, the Ranger standing on Point du Hoc with his rifle is Sergeant Leonard "Bud" Lomell. A historian called Lomell the person most responsible for the success of the D-Day invasion, second only to General Dwight Eisenhower. Lomell searched an inland area for the missing five large costal artillery guns and disabled all five with thermite grenades.

Credit: Larry Selman, artist.

Figure 3. Rangers climbed Point du Hoc.

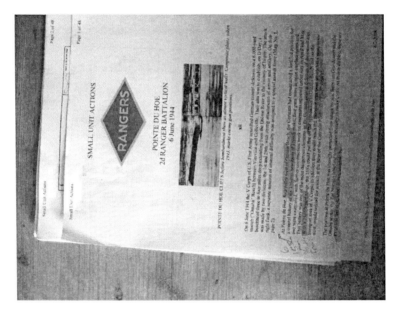

Figure 3.1. Rangers After-Action Report, Point du Hoc.

Chapter 4

SGT LEONARD LOMELL HERO

Leonard "Bud" Lomell lived in Toms River, N. J. and had a law firm in that Ocean County township. Lomell was a friend of mine and accepted my invitation to speak at our Lake Ridge Homeowners Association Memorial Day ceremony.

A meeting room is in the front entrance of Toms River town hall. The room has many of Lomell's memorabilia on the wall. Lomell has many more memorabilia in his home. The first time I visited Lomell in his home, he said, "Charlotte made a cake, what do you want to drink with it, ice tea or whiskey." I knew then that I wanted him to be a speaker in our Lake Ridge Homeowners Association ballroom.

Lomell liked my son, Jay, a U.S. Army colonel. When Lomell and my son started talking, I would go outside to the garden where his wife, Charlotte, was gardening.

Figure 4. Leonard Lomell with author.

Chapter 5

POINTE DU HOC

This photo was Pointe du Hoc in 2014, and it is a popular tourist attraction. You can walk through a few underground fortifications. My son, Colonel Jay Waters, and I visited this point and the many D-Day beaches and cemeteries during a 10-day visit to France.

Figure 5. Pointe du Hoc in 2014.

Chapter 6

SAINTE-MERE-EGLISE

We stayed in Sainte-Mere-Eglise where our paratroopers landed a day before D-Day. Now that town celebrates that event every June. In the photo below, my son Jay and I are seated in a World War II jeep in that town. The driver is a veteran from Holland.

Figure 6. Author and Jay in a World War II jeep.

When Jay was in an airborne brigade that was stationed in Italy, he coordinated the jump of a company of his soldiers into Sainte-Mere-Eglise for a ceremony in which Jay was a speaker on Pointe du Hoc.

Chapter 7

WHITE HOUSE INVITATIONS

I received this plaque for my effort to prevent the move of the Fort Monmouth Research and Development laboratories. I was invited to go to a White House meeting with the Congressional Liaison officer and protested that proposed move to Maryland. Subsequently Congressman James Howard's aide called me. He said as long as Jim Howard is your congressman and Charles Duncan Jr. is the deputy secretary of Defense, the laboratories would not move.

Years earlier I met Congressman Howard. I was the Area Governor of Toastmasters International and wanted some publicity.

I sent a letter to Congressman Howard's office and stated that our Toastmasters International clubs have elected him Monmouth County Man of the Year. We are having a ceremony on a Saturday in June and he and his wife are invited to attend.

He replied that he and his wife are delighted to attend and are waiting for more details. I tried to book the ceremony at the upscale restaurant where we held our meetings. All the Saturdays were booked. I tried other places but the brides had booked all Saturdays in June.

I informed Congressman Howard that we cannot have the ceremony. Then the trophy arrived from HQ Toastmasters International. I contacted the congressman and told him I have a confirmed reservation in July and if he cannot attend I would bring the trophy to his local office in Belmar. He accepted.

Then I invited the two Toastmasters Clubs in the county and the Fort Monmouth Toastmistress. Club. It was a successful ceremony.

I received a second invitation when my daughter Sharon was a volunteer working at the White House. Her supervisor escorted Barbara and me on a special tour of the White House. He let me jog briefly on President Clinton's jogging trail behind the White House. I was also at the podium in the White House briefing room. I "briefed" our group and the technicians working in the rear of the room.

Figure 7.1. "SAVE FORT MONMOUTH" plaque presented to me.

Figure 7.2. U.S. Secret Service mug.

The U.S. Secret Service had a gift shop in the White House that was the size of a closet. I purchased two coffee mugs. I gave one to our chief engineer who had many mugs in his office. Mine is on a high shelf in the kitchen.

Chapter 8

U-2 AIRCRAFT

A Fort Monmouth photographer gave me this photo taken by a U-2 aircraft. On one side of the Garden State Parkway are the R & D laboratories in the Hexagon building and in the Electronic Warfare top-secret building at Fort Monmouth. On the other side of the parkway are all the support personnel who would not move: the production engineers, procurement, human resources, logistics, maintenance, and other support personnel.

The Department of the Army report stated that this is "synergy": The whole is greater than the sum of the parts. Our Save the Fort committee disagreed.

The Army did send representatives to a town hall meeting where local organizations could state their opinions. It was in Eatontown which is adjacent to Fort Monmouth. My friend Tony was the mayor of Little Silver and he appointed me the town Save the Fort representative. This allowed me to speak at the town meeting.

At the town hall meeting I gave various reasons why Fort Monmouth personnel should not move. I showed a display depicting the move of 200 people from the Electronic Warfare Laboratory to Vint Hill Farms, Virginia which had seventy four people. An Army representative asked if he could keep that display. I did mention that the Guardrail aircraft may currently be flying over this meeting. That may have helped the cause.

After the meeting Eatontown Mayor F. told all the Fort Monmouth attendees that all information for Congressman Howard should be given to "Jack." Afterwards a Night Vision Laboratory person told me that they will give their information directly to the congressman. The other laboratories gave give me their information. The Night Vision Laboratory was the only laboratory that moved.

Barbara and I did visit Vint Hill Farms and the surrounding area. We drove around that military installation. There was sign between the road and the fence. It was the boundary line for Prince William County, VA.

That county is part of the National Capital Region and no additional military organizations are allowed to move into the NCR. I gave that information to the congressman and he contacted Vint Hill Farms personnel. Their representative replied that they are in Fauquier County. Sometimes vandals cut down signs and they are replaced in a different area.

Figure 8. Fort Monmouth buildings.

Chapter 9

HIGH- AND LOW-ALTITUDE AIRCRAFT

In Figure 9 the high-flying Mohawk OV-1 surveillance aircraft has my radar jammer on the left wing. The jammer is jamming two radars simultaneously.

Figure 9. Counter-measures against enemy radars.

In the helicopter, my radar locator cockpit display detected an enemy radar signal. Simultaneously the pilot heard the distinctive audio of that type of radar on his headset. He dropped chaff and then made a rapid decent to avoid that enemy radar.

If it was an attack helicopter, the pilot would see the location of that radar by looking at the strobe line on my cockpit display indicator and then fire a missile at it. My radar locator locates that radar before the enemy operator has locked-on to the helicopter.

Chapter 10

MISSILE ALERT SPECIAL TEST

This was a special test to radiate into the blade antenna and measure the frequency range of my missile alert circuit. Note my radar locator antenna on the nose of the aircraft.

Figure 10. Special test with a commercial signal generator.

Chapter 11

FAN AND SPECIAL TEST EQUIPMENT

Another special test was conducted on a lab bench. The hand-held signal transmitter is used on the flight line to test the four spiral antennas and the blade antenna. I made a video of my test. A large overhead pipe made a flushing sound which added some humor to the video.

Figure 11. Special lab test of the radar locator.

The simulator is yellow because a yellow color is required for equipment used on the flight line. My specification for contractors stated that it must be that color in accordance with the U.S. Government MIL-STD-595 standard color chart. The chart was a postage-size picture of actual colors. The non-flight line test items are required to be Forest Green with the associated identifying number in the chart. The radar locator subsystems were a standard gray.

Chapter 12

CANADA BRIEFING

I added a Canadian flag to my material when briefing Canadian personnel. I also briefed German and Israeli representatives, and gave briefings at the U.S. Air Force Academy in Colorado and the U.S. Military Academy in West Point, as well as to U.S. Army personnel in the Pentagon.

Figure 12. Illustration used for Canadian military briefing.

Chapter 13

FIRST INTEGRATED CIRCUIT AT FORT MONMOUTH, N. J.

This radar locator subsystem has transistor circuit boards and one integrated circuit (IC). That IC was the first IC used with a Fort Monmouth production system.

Figure 13. Radar locator subsystem with the cover removed.

Chapter 14

RADAR SIGNAL DETECTOR SET AN/APR-39

I was the project engineer for thousands of these radar locators. Once the black-covered antennas that were installed on fixed-wing aircraft failed. The antenna cover was a very thin plastic and I thought that the combination of wind and rain caused the failure, but I needed to confirm my opinion.

There are many environmental test facilities that conduct rain or wind tests in accordance with Environment Engineering Considerations and Laboratory Tests MIL-STD-810. However, that standard does not include a simultaneous test of wind and rain. I knew of only one test facility that conducted a test using rain as well as wind simultaneously. It was a tunnel chamber at the NASA Ames Research Center (now John H. Glenn Research Center at Lewis Field) near Cleveland, Ohio. So I visited that facility and witnessed the test of one of my antennas. The rain turned to ice, which covered the face of the antenna, so I cancelled all further testing. The rain did damage the antenna, which was the proof that was needed.

I told that information to the system contractor representative of the AN/APR-39 radar locator. Most companies have these reps. They maintain contact with the engineers and contracting officer personnel. They also checked the Request for Bids board in the Procurement Division room.

One rep told me that he knew what was going to go out for bids before it was posted.

I did not get a reply from the contractor about the antenna problem. Tom, another contractor representative, who represented a few electronic component manufacturers did contact me and I explained the problem to him. A couple of days later Tom and the president of an antenna components manufacturer came to my office. They handed me an identical antenna with a thicker cover. My supervisor knew about that president's design of antenna baluns which was in his antenna handbook. They received a contract for the replacement antennas.

That relationship between government personnel and contractors changed over the years. Once the Contractor Representatives Association closed an upscale restaurant for one evening. Government personnel were invited to attend a standup cocktail party that evening. Recommended government personnel received an invitation from the Contractor Representatives Association. My invitation did not state who recommend me and my guest. There were more than a hundred attendees. The commanding general approved this party. He stated that there needs to be more eye-to-eye contact.

Lunch with a contractor also changed over the years. I had many lunches with contractors over the years... Once there was a memo which stated that after having lunch with a contractor you should say, "I will get it next time."

Then things changed. We needed to pay our part of the bill. I paid even though the contactor tried to pay for everything. My production engineer, the contracting officer, and his procurement specialist had lunch at a contractor's country club. They did not pay for anything. The specialist was a friend of the commanding general's daughter. Somehow the general knew what happened and the contracting officer was removed from his job. He stated that there was no opportunity for him to pay. That was not considered a valid reason.

I received 3.1 million dollars to purchase 250 Limited Production models of the AN/APR-39 (V-1). The Procurement Directorate used ten percent of those funds to pay their employees. That is a misappropriation of Government funds. I did not want to complain but I needed those funds. Fortunately the Electronic Warfare Laboratory director gave me those funds from his Discretionary Funds account. One million dollars was reserved to purchase the Government Drawings.

Figure 14. Subsystems of the radar locator AN/APR-39 (V-1).

Chapter 15

TWO RADAR SIGNALS DETECTED AND COUNTERED

This indicator displays two radar signals. The longer thin signal represents a radar at zero degrees which is directly ahead of the aircraft. Hours before the start of Desert Storm, AH-64 Apache and AH-1 Cobra attack helicopters located all the enemy Early Warning radars using my indicator and destroyed them. The enemy Early Warning radar radiated signal flashes ON (strobe line signal) then OFF (no signal) on my Indicator as the enemy radar beam scans the battle area. The other signal on the Indicator which is at 135 degrees is brighter because that radar has a higher pulse rate frequency (PRF) That friendly radar is on our side of the Front Edge of the Battle Area...

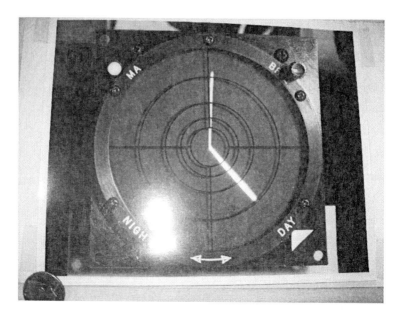

Figure 15.1. Azimuth radar locator in cockpits.

Figure 15.2. The AH-1 Cobra is the top helicopter.

Chapter 16

CONTROL PANEL AND INDICATOR OF AN/APR-39

Both of the subsystems in the figure are installed in the cockpit. When that small white light on the top left of the Indicator is illuminated, it indicates that the enemy fired a radar-guided missile at that aircraft. A special audio signal is also heard on the pilot's headset.

Figure 16. Control panel placed on the indicator.

Chapter 17

WOMAN AND AN/APR-39 PHOTO

I wanted the photographer to highlight my system, not the woman. I placed this system in a briefcase and used it for briefings. If the briefing was at an airbase, I used my battery to turn on the system and we watched as the Indicator displayed and located the Ground Aircraft Control radar signal as it rotated 360 degrees every four seconds.

Figure 17. Radar locator mounted on a plastic sheet.

Chapter 18

FAST PATROL BOAT

A naval officer called me and asked me to install my radar locator on a U.S. Navy SEALs Fast Patrol Boat (FPB). I told him to send me $1,000, and I would come to San Diego and install it on their boat. He said he did not have any money. My supervisor told me to do it anyway.

Figure 18. This FPB is similar to the boat that I installed the radar locator on.

Chapter 19

AN/APR-39 ON NAVY SEALS FPB

This is what I requested the Army Avionics Detachment in Lakehurst Naval Air Station, N. J. to fabricate. The middle subsystem I installed in the Pilot House and the other subsystems I installed on the mast.

Then we left port for a trial run. About two miles off shore the Navy SEAL who was onboard said goodbye to me, then jumped overboard. (At that time there was no GPS or any other device to assist him).

I told a coworker about my project. He told me that he read something similar in the Soldier of Fortune magazine while in his dentist's waiting room. My son Jay had copies of SOF and I found the article.

The CIA had patrol boat in El Salvador used to capture insurgents who traveled in boats which had navigational radars on board. I believed that my radar locator could detect and locate boats using that radar. I wanted to tell that to the CIA.

Previously I attended a Top Secret symposium in a National Security Agency building on Fort Meade, Maryland. I met a CIA employee and had her telephone number. I explained how I could help the CIA. She did not contact me but they may have contacted E Systems Corporation who designed and built the radar locator.

She did tell me that her mother lived in my Little Silver hometown and gave me her mother's address. That house was one that I passed when jogging on one of my routes. So I stopped to talk to the mother who often sat on the front porch. It was a brief conversation. Subsequently when I jogged pass the house I would say "hello" and continued jogging.

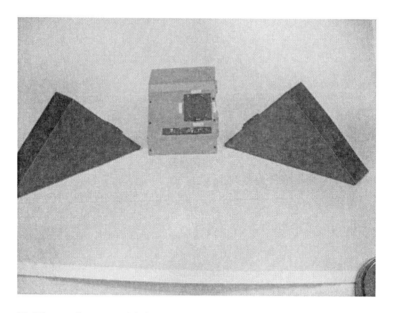

Figure 19. Three subsystems fabricated for use on boats.

Chapter 20

AMERICAN EMBASSY IN IRAN, MISSIONS 1 AND 2

When Ayatollah Khomeini became the ruler of Iran, students attacked the American Embassy. They held many Americans prisoners. We attempted a rescue mission, but two of our aircraft collided in the Iranian desert and the mission was aborted. The World knew about that failed mission, but the World never knew about the second United States rescue mission. I was a technical advisor for the second rescue mission.

At a meeting in Fort Campbell, Kentucky, I briefed the Delta Force pilots about the capability of my AN/APR-39 Radar Locator. I overheard one pilot say, "Thank God for the April 39." We redeployed to California for tests. A crew chief did a Hot Refuel that caused a fire and destroyed the aircraft. I briefed the mission commander about the radars that were in Iran and the capability of my radar locator to detect and locate those radars. Once Ronald Reagan was elected president, the students released the hostages.

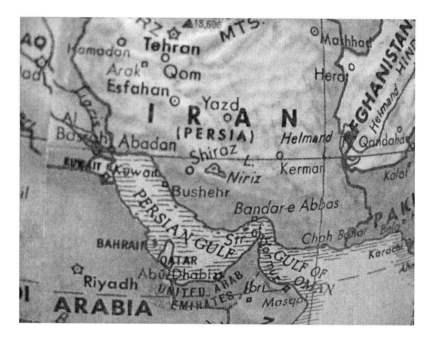

Figure 20. Tehran, Iran.

Chapter 21

BLOCK DIAGRAM OF THE AN/APR-39 (V1)

This is a block diagram of one half of the radar locator system. It locates a radar by measuring the different power received by two adjacent antennas. It will display a strobe line on the Indicator that represents the radar location and relative distance to that radar.

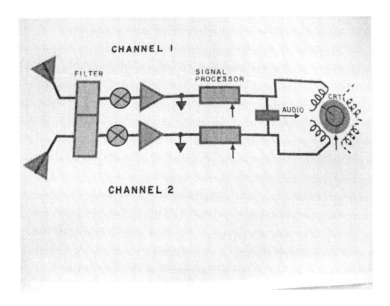

Figure 21. Radar Locator block diagram.

Chapter 22

V1 AND V2 OF THE AN/APR-39

The V2 system displayed in Figure 22 has a digital display rather than the strobe lines of my V1. A number two on the display represents an SA-2 missile guidance system. But the V2 was slow to display an image and sometimes misidentified radars.

When I briefed the pilots, I told them that they will detect the enemy radar before that radar operator detects their aircraft. If the signal from the radar is across the Front Edge of the Battle Area they should assume that it is an enemy radar. If the radar signal is behind the aircraft it is a friendly.

The V1 will also identify the type of enemy radar detected if you listen to the audio signal after having a person explain the difference between audio signals. I knew the difference so when I flew along the North Korea border I detected an SA-2 missile guidance radar signal. When I flew along the East German border I detected a Gundish radar tracking signal. (Our Intel personnel did not know that there was an enemy radar at that location).

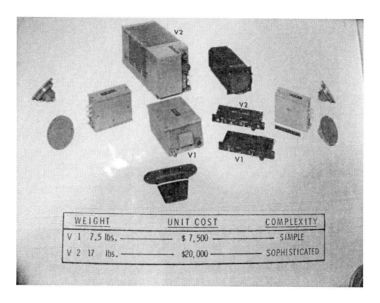

Figure 22. Radar signal detecting set V1 and V2 commingled.

Chapter 23

OH-6 HELICOPTER FLIGHT TEST AND VIETNAM

Figure 23 is a helicopter where I conducted Nap-of-the Earth flight tests over the Chesapeake Bay. The pilot told me that in case of a crash to wait before leaving the helicopter or the rotor blades would cut off my head. Eric also mentioned that he was a POW in North Vietnam. Jane Fonda visited his prison and walked to his cell. She looked in his cell and said to her escorts, "He is OK." Eric was only eating cucumber for the previous six months and wanted to give Fonda --- (The end of the sentence has been redacted).

After the end of the Vietnam War the POWs returned home and reported their torture and starvation. I had seen photos in the New York Times of the POWs playing basketball on an outdoor court and eating in a dining room on plates. I heard statements from my civilian engineering coworkers who traveled to Vietnam for three months as technical advisors. They told me that the Times and other news media were publishing incorrect reports. So I purchased two shares in the New York Times then attended their stockholders meeting in New York City. At the meeting I held a microphone and told Publisher Sutzberger that he was publishing enemy propaganda. The correct reports were in the Daily News. He stated that he does not agree with what I stated and if I left my name and address they would respond in writing. I did receive a letter from the Times editor stating a few errors are made but they try to publish verifiable news. He attached two telegraphs for me from their reporters who were at the battle areas. Also a statement that the Daily News reporter lives and remains in Saigon.

After I attended a second time and spoke the stockholders meetings were held in Arizona.

On our return to Lakehurst Naval Air Station, NJ we flew parallel to the Delaware River. Helicopter pilots use ground features, e.g., highways, train tracks, rivers as navigational aids. Eric asked me if I would like to fly the helicopter. I said teach me and I would like to be the pilot. Eric said to remember WOR (Water on the Right). Then he started working on his paper work. I held the cyclic controller and we continued up river. Then a few antenna towers appeared in front of us that were higher in altitude than our helicopter. I told Eric then he held the controller and we flew around the towers.

Figure 23. Helicopter ready for a flight test.

Chapter 24

SMALL PENCIL FOR PRE-FLIGHT CHECK OUT

In the waiting room at Patuxent River Naval Air Station, Maryland, Eric broke his pencil in half before the first mission. So I broke my pencil in half. Eric said we did that so in case of a crash the pencil would not stab us. Since then I always carry a very small pencil. My coworkers knew about that small pencil and gave me this plaque during my retirement party.

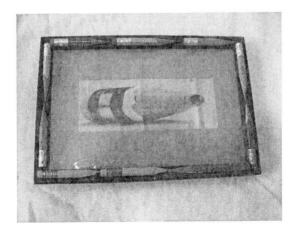

Figure 24. Pencil plaque presented to me.

Chapter 25

CHAFF AND FLARE DISPENSER ON AIRCRAFT

Figure 25. Chaff and flare systems mounted on the side of an aircraft.

Chapter 26

DUAL DISPENSER ON AN AIRCRAFT

A combined Chaff and Flare Dispenser is mounted on an aircraft in Figure 26. When a flare is ejected it travels below the aircraft whereas the chaff travels above the aircraft and blooms. A manufacturer installed chaff dispensers on helicopters in Morocco. Then the Morocco military asked when they should dispense the chaff. The manufacturer told me that he told them that they should install my radar locator.

Figure 26. Combined chaff and flare system mounted on the side of aircraft.

Chapter 27

RUSSIAN GUN DISH RADAR SYSTEM

Figure 27 is the Russian Gun Dish radar directed quad 23-millimeter artillery weapon system. I believed that if World War III started, the Soviets would send thousands of tanks across the Fulda Gap in Germany. They would be led by many Gun Dishes that would shoot down our helicopters. I wanted to have all 2,000 production models of my radar locators installed in all our helicopters before that potential war. My locator would locate a Gun Dish before the Gun Dish operator located the aircraft.

I flew along that Fulda Gap border and located a Russian radar that the Intel people did not know was there. The pilot told me that the month before a young couple tried to cross the border, but the guards shot them. Then guards set up machine guns around the couple and waited. The base commander told his pilots that we are not John Wayne and we are not going to start World War III.

The Army acquired a Gun Dish system and tested it in Fort Hood, Texas. A helicopter pilot flew me to that remote test site. The pilot stated that he had to return to pick up other personnel. He said if I jumped out he would not have to land and do a lengthy Preflight Checkout. I jumped six feet to the ground. My son Jay was in airborne units and jumped many times. I jumped without a parachute.

I also flew along the DMZ in Korea and located an SA-2 Fan Song radar-guided missile system. I asked the pilot to land on a mountaintop so I could get a visual with my binoculars. However the vibrations were so severe that I could not see anything with the binoculars.

During the Cold War, I was told that if World War III started our military would retreat and then set up a defense at the Rhine River. West Germany objected, so we then planned to defend all of West Germany. During the Cold War, all of our presidents supported the nuclear First Strike option. That included the use of strategic ICBMs and tactical one-kiloton nuclear missiles.

Figure 27. Gun Dish deployed in a field.

Chapter 28

CURRENT GUN DISH

This is the replacement for the Gun Dish. This weapon fires missiles, not shells. Note that it has tires.

Figure 28. Current Gun Dish in the field.

Chapter 29

SOVIET SA-7 HEAT-SEEKER MISSILE IN VIETNAM

After the SA-7 heat-seeker missile was introduced during the Vietnam War, our helicopters did not fly in certain areas. Then flare dispensers were added to our helicopters to counter this threat.

Figure 29. SA-7 ready for firing.

Chapter 30

SOVIET SA-2 MISSILE SYSTEM IN VIETNAM

The SA-2 Fan Song missile system has five radars. A vertical scan, a horizontal scan, two target trackers and a missile guidance. When the missile guidance radar (Far right antenna) is activated my missile alert circuit detects that signal and the red light on the indicator is illuminated. Also a distinctive audio alert signal is heard by the pilot.

Figure 30. SA-2 Fan Song is deployed on a mound.

Chapter 31

CRUISE MISSILE JAMMER FOR SPACE AND MISSILE SYSTEMS CENTER AND STRATEGIC DEFENSE INITIATIVE

Two officers from the U.S. Air Force Space and Missile Systems Center arrived at my office. They asked if I would try to jam their new cruise missile. The flight test would take place at the Army Missile Command in Huntsville, Alabama. The test date cannot be changed and no technical information would be provided.

I designed and built two wide-band noise jammers, and a technician and a sergeant drove a military vehicle to the test site with the two jammers. During the journey one jammer malfunctioned. So I worked on it most of the night to repair it. The Missile Command provided a "baby sitter" for me because of the high voltage in the jammer. One jammer was used for the test. I did not have the need-to-know about the effectiveness of the jammer but was told the missile did receive the jamming signal.

That Army truck was modified internally by the Fort Monmouth shops to install a test bench, a table and a desk. It was used for field tests and equipment storage.

I also asked the shop to make a large sign that I could hang the ELECTRONIC WARFARE sign on the side of the truck. Someone saw the sign being made and reported it to the Commanding General. In turn it was reported to the Director of EWL. I was told that I could not deface Government property. The sign was discarded.

The Shops did appreciate my funding them. As the years passed the CECOM Laboratories did fewer internal projects that required tasks that required work by the shops. The projects were replaced by contracts to companies.

I did have problems with the riggers in the Shops who moved heavy items. My AN/ALQ-80 radar jammer weighed 150 pounds. Sometimes when I called the riggers they stated that they were too busy to help me. I had to ask other EWL engineers to help me. Once a rigger saw us moving the jammer and complained to me.

My supervisor stopped in my office. He stated that President Reagan has started the Strategic Defense Initiative (Star Wars) program. If we submit a proposal we could receive a one million dollar contract. I could not think of a relevant project and I was involved with another project.

The High Voltage Pulse Laboratory which was housed in an adjacent building did receive a million dollars for a project. It may have been for a high power Electromagnetic Interference (EMI) transmission device. They conducted some of their tests late at night to minimize electrical interference to the nearby community.

Figure 31.1. S&MSC logo.

I was in a car pool with their Project Engineer. John received the "Army R&D Achievement Award" for his contributions to power modulators. He received an electrical shock on the job and retired early. We remain friends on Facebook.

Figure 31.2. Cruise missile off-loaded.

Figure 31.3. My Field Test Truck.

The SDI program was a factor in ending the Cold War. Millions of dollar were awarded for related programs. The Soviets knew that a successful project could change the strategic defense balance. The Soviets did not have the funding to start a similar program.

Chapter 32

CURRENT RUSSIAN JAMMER

It has tires for greater mobility.

Figure 32. Current Russian Jammer.

Chapter 33

FLIGHT LINE TEST AT LAKEHURST NAS, N. J.

This Flight Line Checkout was conducted in Hanger 5 at the Lakehurst Naval Air Station, N. J. Note the hoods that cover the receiver and transmit antennas of Countermeasures Set AN/ALQ-80. Major M. was my assistant project engineer and helped with the testing.

Figure 33. Simulated flight line checkout.

Chapter 34

COUNTERMEASURES SET AN/ALQ-80

The radar jammer AN/ALQ-80 is seated on its Flight Line manual lift. The jammer connects to the outer-wing pylon. A "break-away" cable harness is connected between the aircraft and the jammer. A fuel tank and the LS-59 strobe light are mounted on other pylons. A Velocity Timed fuse jammer is mounted flush on the other wing.

The radar jammer is on a transport vehicle and is seated on its carrier. There are four pairs of holes in the carrier. When the four poles are placed through the carrier it is a four-man lift. A New Equipment Training sergeant told me that there is a Human Factor requirement that an object cannot be lifted if it exceeds a certain weight. Major M. and I performed a two man lift using two poles and installed the jammer on the aircraft. Therefore lifting is not a Human Factor problem.

View the two hooks on the top of the jammer. They are nineteen inches apart and connect to the outer pylon on the wing. All equipment mounted on the six pylons can be jettison by turning on their dedicated switch mounted in the cockpit. My radar jammer has a break-away electrical cable assembly hooked to a lanyard.

During a surveillance mission along the Demilitarize Zone in Korea the jammer was detached from the pylon and fell to the ground. The pilot stated that he did not activate the switch. A search party was organized but could not locate the jammer. The aircraft was examined and the ejection circuit was operating properly.

Six of my Engineering Development models were manufactured. All the environmental tests were successfully completed, but the Test and Evaluation Command requested that they perform six more months of Reliability Testing. I received a report that the AN/ALQ-80 had successfully jammed an enemy radar. So I called a general in Vietnam. His aide stated that the AN/ALQ-80 was reliable and he would send me a Suitability Statement. I received then sent the statement to higher headquarters. The request for additional testing was denied. Then a Limited Production contract for eighteen jammers was awarded.

One model of the LS-59 strobe light pictured in Figure 34 was being tested one night as I was driving my car. The LS-59 has a bright light that pulses ON then OFF. It may frighten some people. It did spook a horse which tried to jump a fence but became impaled on the fence.

Figure 34.1. Radar jammer on its lift and LS-59.

Countermeasures Set AN/ALQ-80

Figure 34.2. Jammer on its transport carrier.

Figure 34.3. Jammer attached to its white pylon.

Chapter 35

IN-BOARD JAMMER

There are many factors that need to be determined when doing a retrofit to an aircraft. The new center of gravity must be determined and ballast may need to be added to obtain the desired center of gravity.

Prime power requirements need to be determined. An alternator may need to be added. Note that my wing-mounter jammer generates its own air-driven electrical power.

Cooling and Electro Magnetic Interference (EMI) must also be considered. Total additional weight must be determined. Today pilots want Electronic Warfare equipment on their aircraft. Years ago I was the first project engineer to develop an Electronic Warfare system. A pilot told me that every pound added to his aircraft means he has to reduce that amount of fuel on the aircraft. Even after EW systems were accepted by pilots, there was a reluctance to train with it. On a flight line in Germany I asked the pilot why he did not turn on the radar locator. I said that he would need it if there was a military conflict. He said he will be discharged before there is a conflict. I did specify in contracts that AN/APR-39 contractors must have the volume control be designed to have zero audio capability if the pilot wanted to turn off the volume. I did not want a pilot to turn off the radar locator system because he did not want to listen to the audio in peace time.

Figure 35. This jammer was retrofitted to be installed inside the aircraft.

Chapter 36

VELOCITY-TIMED FUSE JAMMER

Figure 36 is a velocity-timed fuse that is screwed onto the tip of an artillery shell. It is designed to explode the shell about thirty feet above the ground. The fuse jammer transmits a signal to t54e fuse so it will explode higher above the ground.

I had a research project called Shortstop, AN/VLQ-9. It was a ground VT Fuse Jammer that was mounted in a High-Mobility Multipurpose Vehicle (Humvee). Its large garbage-can-size antenna protruded above the vehicle. To do a field test, I sat on the floor in the back of that Humvee. A soldier drove over railroad tracks then through some woods in Fort Monmouth as the antenna brushed tree branches. The Shortstop survived that test. Then we conducted tests at Yuma Test Proving Grounds in Arizona and watched as shells exploded harmlessly about four hundred feet above the ground.

Chapter 37

LT. DENNIS W. ZILINSKI II

Currently Barbara and I go to 50th wedding anniversary celebrations. At one celebration a guest gave me a card for the Lt. Dennis W. Zilinski II Memorial Run/Walk. I told that guest this story.

I met Dennis on the Red Bank Community YMCA pool deck. He was a West Point cadet and captain of the swim team. He wanted to be a paratrooper. I told him my son Jay was a paratrooper and a then-Lieutenant Colonel stationed in Italy. I gave Dennis my card and added Jay's name. I told Dennis that if he is assigned to that airborne brigade to introduce himself to Jay.

Then Dennis started to swim laps in a medium-speed lane. Also swimming was Tom Wilkens who held the Olympic record in the breaststroke. Tom was swimming laps in a slow-speed lane. I swam my laps in a fast-speed lane. When I told Tom's aunt how that happened, she laughed but Tom was not amused.

After Dennis graduated, he was engaged to his classmate sweetheart and they both were deployed to Iraq with the 101st Air Assault Division. Dennis was killed in Iraq when his vehicle was hit from an improvised explosive device (IED).

Every year there is the Lt. Dennis W. Zilinski II Memorial Run/Walk event in the Garden State Art Center area. I walked the two miles. I once completed the New York City marathon, but my running days are over. The West Point swim team did do the run, as well as the Commanding General of the 101st Air Assault division. There I met Dennis's mother and told her the pool deck story.

Figure 36.1. The West Point swim team before their two-mile run.

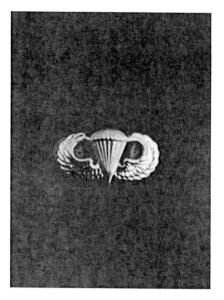

Figure 36.2. Parachutist badge.

Chapter 38

TOP SECRET MEETING AT THE SENSITIVE COMPARTMENT INFORMATION FACILITY

After an Intel officer called me at my office, I would drive to their SCIF (Sensitive Compartment Information Facility). There usually would be new information about a Soviet weapon system. In addition, there would be new information about many other subjects. Some would be reports of current events, which would be different from what I was reading in the news media. I tried not to read them because I was concerned that I may say something classified that others would repeat.

There was a report that a high-ranking Soviet general was seen in the Moscow railroad station on a certain date. That is Top Secret because of the source. Who was there who saw the general? Soviet KGB analysts may be able to identify that person who is providing information to the United States.

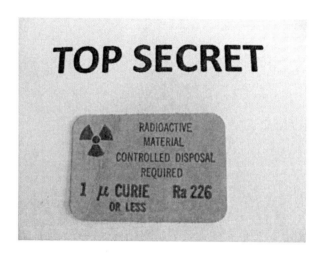

Figure 37. Top Secret.

Chapter 39

FIREFINDER AN/TPQ-36 IN AFGHANISTAN

Figure 39 is the mortar locator, AN/TPQ-36 Firefinder. When Jay was stationed at Bagram Air Base in Afghanistan, that system was used to locate mortar sites. However the insurgents would fire a mortar near a school or mosque so we did not return fire.

Figure 38. This is the AN/TPQ-36 behind a defilade.

Jay was the Adjutant General member on the commanding general's staff. The general summoned the town leaders to a meeting in his office. He told them he wanted the insurgents who are firing the mortars. If they did not bring them to him, he would closed the market outside the base. That was the local people's money-maker so the town leaders did bring two men to the general. Jay told me that they may not have been the insurgents who fired the mortars, but the shelling stopped.

Chapter 40

FIREFINDER AN/TPQ-37

This is the AN/TPQ-37 Firefinder. The latest Firefinder is designed to detect and track incoming artillery and rocket fire to determine their point of origin so we could initiate counter-battery fire.

Reportedly if an artillery shell was fired thirty miles towards Fort Monmouth from a building-top in Newark, then the Firefinder would locate the firing site. Then our artillery could return fire before that shell hit its target. Coriolis, the Earth's rotation, was factored into the equipment's accurate firing solution.

I was a consulting engineer during the Firefinder system development phase. The contractors who would bid on the Firefinder systems stated that they could not meet the accuracy requirement. The government engineers stated that the accuracy is required. I recommended that at close range the requirement must be accomplished. But the long range accuracy could be less. The government engineers agreed and they reached an agreement with the contractors as to what range change would be permitted.

I was asked to review the winning proposal. I reviewed the parts list cost then contacted some of the vendors to send me a price quote. All the quotes were less than what the contractor quoted. I believed that the contractor submitted the highest cost from vendors then would order from the lowest vendors. I submitted the quotes that I received to the government engineer.

Lieutenant Colonel C was in charge of the Firefinder program. He told me that his superior asked him to find a way to save money then he asked me to find the way to save the money.

The Firefinder has a backup electrical generator because of the high failure rate of the generator. However that second generator is not deployed in combat areas. I was aware that the Military Generator Program Manager in Maryland had a new generator with a high Mean Time Before Failure. I scheduled a meeting then traveled to that office. As a result of that meeting I told LTC C that he should consider using that generator and save money by not having a second generator. Then the Firefinder Production Engineer stated that the second generator was needed and the subject was tabled.

LTC C. then asked me to review their budget. He told me that, "I could do anything." I did do it without working with the budget manager.

Then LTC C told me that he was selected to attend a school which would advance his career and he was leaving. So it was also time for me to leave and I returned to the Electronic Warfare Laboratory.

Figure 39. AN/TPQ-47 deployed in a field.

Chapter 41

OLD CROW FIGURINE

When I was the President of the Garden State chapter of the Association of Old Crows, we had guest speakers. I would give each speaker an Old Crow figurine, and a Fort Monmouth photographer would take a picture of me giving the figurine to the speaker. Then I sent that photo and a meeting report to the editor of the Old Crows' international publication. After the third submission the editor told me that they would no longer publish that figurine, because it contains whiskey.

As previously mentioned, the Crows had a yearly national convention that was usually in the Hilton Hotel ballroom in Washington, D.C., and congressmen would be the speakers. Once they had the convention in Dallas, Texas, so I asked Barbara to come with me. We had never left our three young children overnight so Barbara was concerned. A road-runner friend recommended a woman who rented a home that he owned.

Then I purchased a TV for all of them to watch. Until then I did not want a TV in our home. Our son Jay also did not want a TV in his home, when he was an adult. Then his father-in-law, Sergio, arrived from Italy to stay for a week, so Jay purchased a TV. Sergio wanted to watch the Olympics.

Figure 40.1. Old Crow figurine.

Figure 40.2. Back of Old Crow figurine.

Chapter 42

COLD-WEATHER TEST ALASKA

When we required our contactors to perform environmental tests, we added appropriate tests in accordance with MIL-STD- 810 to the contract. The cold-weather test was conducted at a temperature of minus 55 degrees centigrade (-67 F).

Another project engineer wanted to conduct a separate outdoor cold-weather test. He told me that Fort Drum Army base, N.Y., was the coldest place in the United States. (He meant the contiguous states). Fort Greely, Alaska, is the Cold Regions Test Center and one of the coldest areas in Alaska.

Fort Wainwright, Alaska, is close to Fort Greely. I briefed their pilots from the Aviation Brigade Rescue Mission for an Army hospital. When I arrived for my briefing, I was told that it was the start of moose season so half of the pilots were not there.

Many of the soldiers considered that good duty, but other soldiers or their wives had Seasonal Affective Disorder. The base commander did not like that command. He told me that his aircraft were painted red, and he thinks that he commands a circus.

During World War II, Fort Wainwright was called Ladd Army Airfield. About 8,000 military aircraft landed there. Russians pilots received two weeks training at the airfield then they flew the aircraft to

Russia. The pilots who delivered the aircraft were women from the Women's Auxiliary Ferrying Squadron (WAFS).

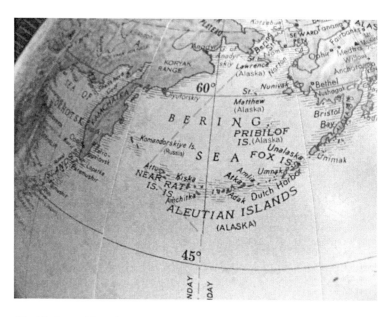

Figure 41. Alaska and Russia.

Chapter 43

SLINGSHOT

As a teenager, Jay decided that he wanted a military career as an Army officer so he received an ROTC scholarship from Wake Forest University, NC. I did not know why he decided to join the military. He never had a toy pistol, rifle or knife, but he did have a slingshot. A young boy from around the corner would come to the front of our home and pee on our ivy. Once Jay opened the front door, then aimed his slingshot at the boy.

His father was a state trooper and the next day a Little Silver police officer arrived. He came into our home and told us that the trooper had filed a complaint. I told Jay to give his slingshot to the police officer.

Jay may have decided on a military career the day Barbara signed him out of grammar school. A helicopter from the Lakehurst Naval Air station flew up to Fort Monmouth to pick me up. I would be conducting a flight test over the Chesapeake Bay. A Fort Monmouth photographer was there to take photos of my onboard radar locator. He also snapped a photo of Jay standing in front of the helicopter holding my helmet. The photographer gave me a copy of that picture. I cannot locate that photo, and Jay stated that he does not have it.

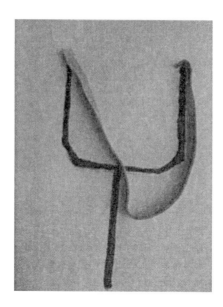

Figure 42. Slingshot.

Chapter 44

UNDERCOVER AGENT FOR
A GOVERNMENT AGENCY

My supervisor Ray stopped in my office. He wanted me to help a federal crimes agency. A contractor was trying to sell radar jammers to a foreign country. Contractors are required to get approval from the foreign military sales agency. (A couple of times that agency called me for my recommendation for a sale to a foreign country).

I traveled to the agency to determine how I could assist them. Jerry stated that I would be the representative of a Middle Eastern country to arrange a sale of a radar jammer. That Middle Eastern country representative (Jerry's co-worker) would arrive by air at a local New Jersey airport to meet with me and the contractor. A sale would be negotiated, then they would arrest the contractor.

I told Jerry that during my first telephone call to the contractor I would mention that I have my design for a radar jammer. Then Jerry arranged for me to meet with their legal team at a different location. A lawyer told me that I must not mention my design. It could be considered collusion.

Jerry stated that they would pay me $1,000 for every week that I worked. (I would have done it gratis). There was a remote chance of me being in physical danger; if so, they would put me in the U.S. Federal

Witness Protection Program. Then I made a call to the contractor, but he was not in. Jerry stated that I said all the right things.

Jerry gave me the bio of the contractor. It listed one published paper. I located that publication and noticed that there was a similar subject by a different author in Italy. After reading both papers, I knew that the contractor had plagiarized all the technical information from the other publication.

Jerry gave me my business card. I was the president of J & B Consulting Company, and my home telephone was on it. The fax was in Jerry's office. When Jerry would call my home, Barbara would say that I was not at home. She knew that it was Jerry because of his deep voice and he only said, "Thank you."

They decided that it would be better to have the contractor call me at my office. I shared my phone with another engineer who was in a different room. I told Bruce that an unknown person may call me, and if I was not in that he should only say, "I was not in." Someday I would tell him what was going on. If he was very concerned about this unusual procedure, he could talk to Ray.

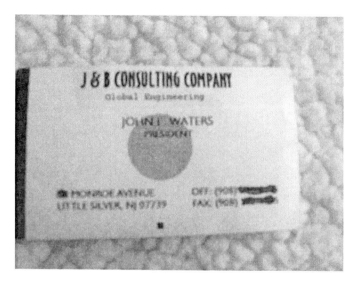

Figure 43. Business card provided to me.

Undercover Agent for a Government Agency 97

Ray told our laboratory director about this project. The EWL director said we must tell the commanding general because he needs to know what is going on in his command.

The agency decided that too many people would be aware of this project so they cancelled my involvement. I had mailed Jerry invoices for two weeks of work. I never received payment, so I called Jerry. He said someone else would get the money to pay me. He was being promoted and going to Washington, D.C. I sent a letter to the director of the agency. Then Jerry called me and asked me to come to his office to be paid. He gave me the cash while a co-worker watched but he never he said a word.

Chapter 45

VIETNAM AND THAILAND

During the Vietnam War, the U.S. Air Force deployed a new Radar Warning Receiver, AN/ALR-46. We needed to know if it was suitable for our Army fixed-wing aircraft. I was tasked to go to Thailand to visit a U.S. Air Force base that had those Radar Warning Receivers installed in their aircraft. I requested that Jim, my reliability engineer, accompany me.

We traveled on a Military Airlift Command (MAC) flight out of Travis Air Force Base, California, to Clark Air Base in the Philippines. We remained overnight in the adjacent Angeles City. I discovered where all the jeeps in WW II were relocated. They are in that city, but all are painted in bright colors.

The next MAC flight destination was to Bangkok, Thailand. The first flight had all military personnel. On this flight we were the only passengers. It was the "Stars and Stripes" newpaper delivery aircraft.

Jim and I spent 10 days in Thailand going to different U.S. Army air bases. Because of security concerns we shared the same room, ate together, went to meetings together and shopped together. After six days I disliked being with him, so we shopped and ate alone. Jim was interesting and once arrived at a restaurant in a gin rickshaw. When he left the restauant he told the gin to sit in the rickshaw, then Jim pulled it back to the hotel.

We stopped at one base that had a very large antenna. The Executive Officer told me that there was one intercept that justified having the base. That major told me that they built a lumber mill then gave it to the town. (Give a man a fish and you feed him for a day. Teach the man to fish and you feed him for a lifetime).

At a U.S. Air Force base In Northeast Thailand, the Army had fixed-wing aircraf flying sorties in or near enemy territory. Some had an AN/APR-25 radar warning receiver on the aircraft. An aircraft without the AN/APR-25 was shot down. So the commander stated that aircraft without the AN/APR-25 must to be acompanied with an aircraft that had one on board. I tested aircraft that was on the flight line. One had a defective AN/APR-25 warning receiver subsystem. I borrowed a screwdriver from a contractor to remove that subsystem. The screwdriver was "lost" so I went to the PX and purchased another screwdriver for him.

At the airbase were USAF F-4s on the ready flight line with the latest AN/ALR-46. They were not flying combat missions. The Army was flying combat missions using the older AN/APR-25. I arranged a meeting between the Army and Air Force pilots. Jim examined the reliablity reports for the AN/ALR-46 at the USAF base.

Figure 44. Rickshaw (photo permission from iStock by Getty Images).

Vietnam and Thailand 101

On our return fligh we stoped at Saigon, Vietnam to refuel. The pilot said that we could not leave the aircraft. Our Government stated that there were no Americians in Vietnam. If we left the aircraft, we could be on the front page of, "The New York Times."

Chapter 46

JOINT SPECIAL OPERATIONS COMMAND AND UAVS AND SATELLITE TRANSCEIVERS

Whittaker Corporation in Simi Valley, California, was the manufacturer of the Shortstop fuse jammer. The company also built a radar system that is used at field test sites. The Joint Special Operations Command (JSOC) was interested in purchasing a radar for one of their test sites.

Colonel Rick from JSOC and I visited Whittaker to discuss the operation and purchase of that radar. During one of our early-morning runs, I asked Rick if I could go on a Special Ops mission as a technical advisor. He said, "No way."

Months later, Rick called me. I should put in travel orders to go to Fort Benning, Georgia. I was not to tell anyone where I went or what I saw. I watched an airborne training mission. I cannot reveal anything about it because I have never heard or read what happened at that type of mission so it may still be classified.

JSOC was also interested providing the LST-5 SATCOM Satellite Transceiver for a special customer (a Black Program). JSOC purchased two of the first production transceivers. (Secretary of State Colin Powell received the first transceiver). There was a meeting in Washington, DC that included Naval Research Laboratory personnel. I acquired a special container to house the transceiver. The customer asked if we could design and build a physically different type of antenna that could be installed inside that special container. I remember my supervisor stating that we "may" be able to do it.

I hired a known antenna expert who recently retired to help me. We would build a spiral antenna. I asked the sheet metal shop in my town to build it. I told him our military needs it. He did build it but stated that it was difficult to do so he would not build anymore.

We tested the antenna in an anechoic chamber in Fort Monmouth. Then the customer scheduled a high priority time period for us to contact the satellite. We did transmit and received a signal. However we were not sure if the signal was from the satellite or the second transceiver which was on the other side of a big building. Then we transmitted a signal during a field test in Maryland that was observed by the customer. We did not received a signal. The antenna did fit in the container but it was too small for that satellite frequency. It could receive that satellite signal if it had sufficient radiated power. It did not. The Naval Research Laboratory was given all of the program that we were sharing.

Working with transceivers was interesting and the detecting and jamming of our communications is currently a national problem.

There is a way to detect jamming from hostile nations. Currently all transmitters and computers messages are transmitted as a "One" or a "Zero" bit. (A high pulse or no pulse). Many messages are eight bits in length representing a letter or number. If any of those eight bits contain one or more Ones which should be Zeros the message was jammed.

Rick's next assignment was as a battalion commander with the 101st Air Mobile Division in Fort Campbell, Kentucky. There, he tested Unmanned Aerial Vehicles (UAV). He gave me a JSOC memento with my name on it.

A 60-pound UAV was one of the UAVs that Rick was evaluating. That UAV and the Pointer UAV were the two that I thought Fort Monmouth should support and fund.

At that time I was a consulting engineer for the Fort Monmouth UAV Project Office. One of their engineers and I traveled to Washington to attend a UAV conference. That evening we had dinner with a major general. He stated that we should have a, "build-a-little, test-a-little" UAV program. That is what I put in my trip report. The project supervisor liked that report and sent it to higher command. The Project Office did not state which UAV they would, "build a little," so it was not funded.

Then that project office had a meeting and invited many of the majors and lieutenant colonels at Fort Monmouth to attend. The officers wanted to be able to, "see over the hill." They did not recommend what UAV they wanted. I mentioned that war has been described as, "long periods of boredom with short periods of terror." If they selected the less than 20-pound Pointer UAV, the force could start the, "over the hill" observation missions immediately. The Pointer is slow speed and has a limited field-of-view of the battle area. But there should still be enough time to reconnoiter the entire area and continue to repeat the search area. It is a quiet and a low observable UAV.

I could tell that the military did not like that quote and they did not like a civilian contractor recommending a UAV. Again, no recommendation was made and Fort Monmouth did not have a funded UAV program. The Pointer is currently used by Special Operations Forces and many other Forces.

Figure 45.1. Launch of the Pointer.

Figure 45.2. JSOC memento.

Chapter 47

GERMANY AND GERMANS

I did speak to under graduate students from Monmouth University in Long Branch, N. J. about the Cold War. I was also asked to speak about Germany and the German people.

When I first traveled to Germany, a liaison officer from the U.S. Army Aircraft Survivability Office in St. Louis, Missouri, greeted me. He told me that he once was stationed in France. Then Charles de Gaulle became the president of France and he "kicked out all of the Americans." The liaison officer liked France and his neighbors liked him. A week after he moved to Germany, his new neighbors knocked on his door and said, "Mein Herr, you did not sweep your sidewalk." They keep a cow in the house, but they do have flowers in pots on the windowsills.

My first official contact with a German was when I went to an on-base office to buy gas stamps that are used for a lower price at a gas station. In error, I purchased stamps for high-test gas. I told the clerk I needed regular gas stamps. He said that would be two transactions. He gave me back my money for the high-test gas. When I asked to buy the stamps for the regular gas, he told me I had my gas limit for that month.

Figure 46. Rommel the Desert Fox War Prize.

Field Marshal Erwin Rommel and his Afrika Korps discovered a huge wine and spirits cellar during one of their battles in Africa. After disbursing some of that war prize to the soldiers, it was shipped to Italy and then to the American military in Germany. It was disbursed to our military for special occasions. The German government became aware of this and stated that all the wine and spirits belonged to them. So they went to court to get it. The USASEUR Class VI Council decided to sell it to our military at all bases, pending a decision from the court. I purchased two bottles of scotch and kept them unopened until this year. My grandson Albert collects alcoholic bottles, so I gave him one bottle.

Chapter 48

NO SPARE PARTS OR AIRCRAFT FUEL

I was a member of the Program Manager Aircraft Survivability Equipment team that visited aircraft maintenance facilities in West Germany. At an air base, I was with the maintenance specialist in his General Support shelter. I looked at his spare parts inventory. There was a very large part that was not a component of my Radar Signal Detecting Set AN/APR-39 (V-1). The tech stated that the part is listed in the APR-39 parts manual, and he ordered it. The shelter door opened, and General Patton's son, Major General George Patton IV, entered. He left without making eye contact with me. The general was the commanding general of that division. At that time there were four divisions and two corps in Germany.

I never reviewed that P-manual, but I did review the draft Operators' manual. It was poorly written. I contacted my manual's technician and complained. He said he only checks for format. Then I called the contractor. He stated their contract requires that the manual be prepared and delivered before any APR-39 models are built. I rewrote more than half of that manual.

Note in Figure 48, AN/APR-39 disassembled parts, that some subsystems contain printed circuit boards. At the general support level, troubleshooting only requires locating the defective PC board and then replacing it. The defective PC board is sent to the depot (Tobyhanna Army Depot, Tobyhanna, Pennsylvania).

To manufacture a PC Board, the transistor leads are inserted in the designate holes in the board then it goes into a wave solder machine. The solder contain lead and lead may no longer be used. My plumber told me that plumbers no longer use solder which is half lead.

During a contract proposal review, there was a problem about the PC boards so I called the contractor for more information. Then I received a phone call from the Pentagon. The caller said the acting Secretary of the Army wants to know why I am delaying the contract award. I said the incident was minor and would only delay the award a couple of days.

I made a Note for File about that FONECON then gave a copy to the Electronic Warfare Laboratory director. He thanked me and stated that he could use what I gave him. I also sent the note to the contracting officer. Months later the incumbent contractor won the award but another bidder complained to his congressman and filed a complaint. I previously paid $1 million dollars to the contractor for the federal government drawings so we did not need to have a sole-source contract. That contractor told me that they needed to win the multiple-source contract, and they left $2 million on the table to get it. (The contractor hoped to make a profit from military sales to other countries).

The General Accounting Office (GAO) of Congress did investigate. The contract specialist called me and asked about the note I sent them. I told him to leave it in his file. I also left the note in my file. Two GAO officials came into my office and looked at my files, but they never saw the note.

Our commanding general needed to testify before a congressional committee about the award. He stated that there is no evidence of a buy-in. He said he would monitor the contract very closely and watch for potential overruns. There were no overruns, and the general told me he wished all of his contracts could be like mine.

No Spare Parts or Aircraft Fuel 111

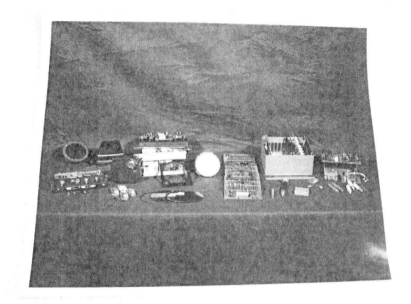

Figure 47. AN/APR-39 disassembled parts.

The Aircraft Survivability Equipment team did go to an airbase that had Mohawk OV-1 surveillance aircraft. The commander stated that everything was fine. During our conference he realized that we were not inspectors and we might be able to help him. The major stated that he did not have any JP4 fuel that is needed to fly his aircraft. Many of the air bases had one or more Hangar Queens. Parts were removed from them and then installed in other aircraft.

After President Reagan was inaugurated, he became aware of the shortage of spare parts and fuel, and authorized contract awards to solve that problem.

Chapter 49

FORT RUCKER COMPLAINT

A pilot from Fort Rucker in Alabama complained that my radar locator had a deficiency. When his helicopter descended to the nap of the Earth, the radar locator lost the signal but a radar continued to track the helicopter.

Figure 48. Antennas installed above the rotor blades.

Fort Rucker housed the Army Aviation Center and School, so action had to be taken to address the problem. I had the four spiral antennas mounted above the rotor blades. That would solve any Nap of the Earth problem, but there could be many unknown problems such as accuracy, mechanical malfunctions, etc. I flew to Fort Rucker and met with the pilot filing the complaint. He agreed to fly me in the helicopter and show me the problem. There was no problem. The radar broke lock on our aircraft, but my onboard radar locator continued to detect and locate the radar. The complaint was withdrawn.

Chapter 50

NAVIGATION AID FOR EMERGENCIES

Rocky was a project engineer in the Avionics Laboratory at Fort Monmouth. He told me that he was in a helicopter in Belgium when a thick fog covered the aircraft. The pilot could not see his visual markers, which he needed to return to the airport. Rocky told the pilot to turn ON the switch for the radar locator, then turn the aircraft so the strobe line on the indicator was at zero degrees. If he maintained the strobe line at zero degrees, then he would arrive at the airport.

Rocky suggested that I recommend the radar locator as a navigation aid. I said the Avionics office would not approve it. Pilots will learn how to do it if they turn on my radar locator when they fly. I asked a pilot why he did not turn the radar locator ON. It may save his life during a conflict. He said he will retire before there is a conflict. My system contract specified that the volume control must be able to be turned to zero volume. I did not want the pilot to turn the radar locator to OFF because the sound was a distraction.

A pilot told me he and another pilot in a second helicopter triangulated on a ground radar and determined the distance from the radar to their aircraft. My system only determine relative distance. One ring on the Azimuth Indicator indicates that the radar is far away from the aircraft. Four rings indicates that the radar is close to the aircraft.

Figure 49. Power (PWR) switch is ON and the airport radar is directly ahead of the aircraft at zero degrees.

I did not recommend pilots be trained to do triangulation. If there was a need from a field commander or the Training and Doctrine Command I would assist as needed.

Chapter 51

ITALIAN FASCISM

My son COL Jay gave me a bottle of wine that he purchased in Italy when he visited his wife Anna's parents. The bottle has a bundle of fiber strands around its neck. Jay explained its meaning. Fasces is the symbol of a bundle of sticks that represents power in ancient Rome. During World War II, the Fascists and their dictator Benito Mussolini used fasces as their rally slogan "One fiber is weak but we are many fibers and we are strong."

Figure 50. A bundle of fiber strands.

Chapter 52

VIETNAM URGENCY

A major from the Aircraft Survivability Equipment Program Manager's office in Saint Louis, Missouri called. He said that our country is pulling out all of our military in Vietnam. We are leaving 400 helicopters with the Army of the Republic of Vietnam (ARVIN) and they need to have radar detectors installed.

American Electronics Laboratory (AEL) personnel demonstrated to the PM-ASE personnel a radar detector that will detect the Gun Dish radar. PM-ASE wants me to start a program to deliver 400 of those detectors within thirty days.

I received one model of the locator. It was the size of a slice of bread. It could fit on the inspection plate that was on the front of a helicopter. I planned to direct AEL to deliver all locators mounted on that size inspection plate. ARVIN military could remove the plate currently on the aircraft then install the plate with the locator.

The locator would have a six wire cable assembly. Two wires would be connected to the battery and two would be connected to the aircraft audio circuit. The other two would be connected to the RECEIVERS module located in the cockpit. Note that the installation of my AN/APR-39 system in a fixed wing aircraft required 200 man-hours. This installation would require an estimated five hours.

Figure 51.1. Helicopter with inspection plate highlighted.

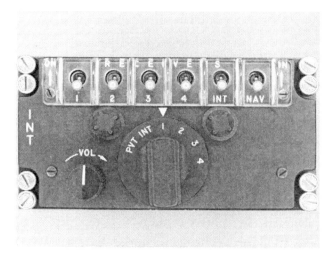

Figure 51.2. Receivers module which is mounted in the cockpit.

The Avionics Laboratory had a similar requirement for an Avionics system. So an Avionics Lab team and I flew to the Air National Guard, Gulfport Combat Readiness Training Center, Mississippi.

Vietnam Urgency 121

I completed my installation there. The Avionics Lab installation required removing metal pieces of the aircraft. The Major in charge of the center caution that team not to remove any load bearing beams.

Then the aircraft and I travelled to Eglin Air Force Base, FL to test the installation against a radar. The pilot told me that due to low cloud cover we could not fly that day. I told the pilot about the program and that I could not delay a day. He agreed to fly the mission. After taking off we were in fog so dense that I could not see anything. I asked him to return to the airport.

I was able to perform a different test. I once purchased ten Navy yellow flight line flashlights. I removed the bulb then installed a relay. The relay had a loud chatter and transmitted a noise frequency signal. I used it to make a quick test of the AN/APR-39 four spiral antennas. That frequency noise signal was also detected by the new radar detector that was installed in the aircraft. I called the relay manufacturer to purchase nine more relays. I planned to give those modified flash lights to the ARVIN military. The manufacturer's sales representative told me that the relay was replaced because it was too noisy and loud. The sales rep said my request was funny and he was going to tell his president why I wanted it.

In order to produce 400 locators AEL would request their competitor, Loral, to produce half of them. I called the major at PM-ASE and stated that we will do what he requested. Both AEL and Loral needed to be funded. Days later he returned my call. The program was cancelled. I did not ask why.

Chapter 53

NEW RADAR SYSTEM

The United Kingdom had a radar system that was higher in frequency than the bandwidth of the detectors in the AN/APR-39 Receivers. The Program Manager for Aircraft Survivability Equipment requested my comments. I stated that I will remove the two bandpass filters from the forward dual receiver of a UH-1 Iroquois (Huey) helicopter at an England test site. That AN/APR-39 may be capable of detecting the UK radar.

PM-ASE arranged a flight test in England. The team consisted of our leader from PM-ASE, two Warrant Officers who would fly the rotary wing aircraft, two Army Captains who are fixed wing aircraft pilots from the Aviation Center in Fort Rucker, AL, and myself.

We stayed in the Officers Club at the UK Aviation Center which is about 80 miles West of London. Warrant Officers are not allowed to stay in the OC but some arrangement was made although I never saw them in that building. At night we left our shoes outside our door. The orderly would shine them then he brought me my shoes and a cup of tea while I was in bed. I told Barbara that I like having tea in bed. She said that will not happen in our house.

At dinner the table setting had nine utensils. I knew you start using the outside utensils first but it was confusing. Our granddaughter Sarah was in Norway as a Student Exchange student. At her host dinner table they had many utensils and lighted candles. The next time Sarah came to dinner at our house I had Teacup LED candles and gave one to Sarah and her mother Anna. I then purchased more candles at the Dollar Tree store to be used when we have guests, etc. Magic is my hobby so I would produce illuminated candles using magic then give one to each female guest.

We had a free weekend so I went to Ireland. My dad would have wanted me to do it. I boarded a ferry in Wales then traveled by train to Dublin. In Dublin there was a worker in the street using a jackhammer and wearing a Harris Tweed jacket. I made up a story that I told many times.

When an Irish boy has his First Holy Communion he is given an oversized Harris Tweed jacket. He wears it to his confirmation, marriage and funeral.

Our path to the test site required a helicopter flying over Stonehenge. I closely examined that layout of stones. It may be a make-work project in order to keep their troops busy doing something.

At the test site the radar tracked the helicopter and my radar locator received and displayed a strong strobe line in the direction of the radar. It was a successful field test. Later after we landed and were in an open field when a Norwegian fighter aircraft buzzed us flying about one hundred feet above ground. That aircraft was there for a different field test.

While on the ground and I was in the helicopter so I asked the pilot to turn ON the radar locator then start the engine. The radar locator malfunctioned. A high transit voltage burnt out the voltage regulator diode. I removed the Control subsystem from the aircraft and replaced the diode in the maintenance shack.

I believed that the reason for reported failures is because pilots do not turn OFF the radar locator power ON/OFF switch. Now I had proof. There is a maintenance bulletin published monthly. It is a pocket size cartoon type bulletin that report failures and fixes. I submitted my failure/fix report to the publication.

There is a diode that has a higher rating and will survive the high voltage transits. It is a non-standard component so the contractor did not use it. I put in a request to have that diode qualified for use in military systems.

New Radar System

Figure 52.1. Table setting at the Officers Club. (My table).

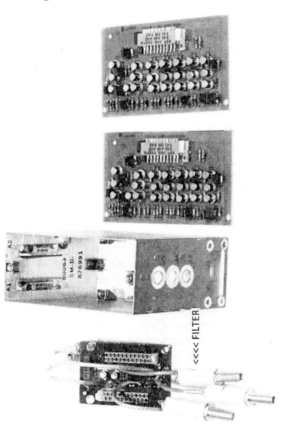

Figure 52.2. Two filters were removed from a receiver.

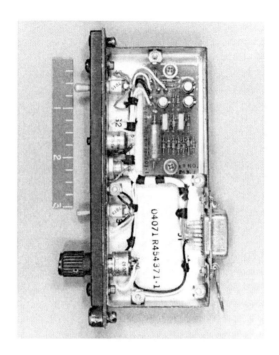

Figure 52.3. The Control contained a defective diode.

Chapter 54

VELOCITY TIMED FUSE LIVE TEST

The Shortstop is a Velocity Time fuse jammer. I received a couple of VT fuses from the Harry Diamond Laboratories, MD and used them for bench tests.

My supervisor returned from Kuwait after Desert Storm ended. He asked me to obtain some VT fuses that were scattered around the Kuwait desert. He gave me a point of contact and I sent an email to that major. I also sent a copy to a colonel in the Special Forces Command. The major replied that it was a problem to obtain the fuses. The colonel then sent an email stating the major should solve the problem and do it. The Foreign Intel Branch found out what I was doing and stated that it was their responsibility. They did not involve the colonel and the fuses were never shipped.

A colonel was coming to our laboratory to discuss Shortstop and other projects. My supervisor told me that they wanted a realistic outdoor test. I asked if they wanted me to use gunpowder. He did not reply and left my office.

I visited a sportsman store. They had a barrel of black gun powder near the checkout counter without a top on it! I purchased the amount that the clerk recommended. I asked the Explosion Ordinance Detachment (EOD) to assist me. (I met their two sergeants when I visited their library to do research on fuses). The test was a booming success.

Chapter 55

HONDURAS BRIEFING

The Electronic Warfare Laboratory received a request from our military in Honduras for technical assistance regarding my radar locator that was installed on their fixed wing aircraft. I was requested to provide the assistance.

There was anti-government unrest in Honduras so I requested that the American Embassy have a person who will meet me at the Tegucigalpa airport. I carried my radar locator which was packed in a briefcase. That could be a problem with security at the two airports. So I prepared a letter under our laboratory letterhead and stated that if there was any concern to contact the director. The director signed it and added his telephone number. At the Newark NJ airport boarding area I was requested to explain what was in the briefcase. My explanation was accepted. After boarding the aircraft the cockpit cabin door was open so I showed the pilots. One said he was familiar with warning receivers.

At the Tegucigalpa airport there was no inspection, you deplane and walk to your transportation. Someone from the embassy was there and escorted me to my hotel. There was mob unrest a couple of blocks from the hotel. I put a piece of tape on the hood of the car. If it was removed I would not start the car. I ate all my meals in the hotel.

The air base was about a distance of thirty miles to the North. I located the road I wanted and started driving. There were no road signs once I was outside the city. I would choose what I though was the major road and drove on it. (In Germany there are signs that indicate which is the major road). At the base I was greeted by a civilian contractor technician. We discussed the problem then walked to the flight line. George watched as I performed a Flight Line Check Out on all the aircraft. One aircraft had a defective Receiver. George stated that he would replace that Receiver.

I briefed the pilots. They fly high and slow and are easy target for enemy radar directed weapons. I recommended that if they hear a radar locator audio tone to look at my radar locator indicator for the direction of the radar. Then take appropriate action. I stated that a one ring strobe on the indicator indicates that the radar is far away from the aircraft. If the strobe line continue to gets longer the aircraft will eventually be in the lethal range of the weapon.

Figure 53. OV-1 Mohawk aircraft which has my radar locator and radar jammer.

Chapter 56

PANAMA CANAL ZONE BRIEFING

Jack from PM-ASE and I visited the HQ US Southern Command. We landed at the Panama City airport. Braniff International Airways lost my luggage. At the hotel I washed my undershirt but in the morning it was still wet. I left a sign in the room stating that I will return. The cleaning woman did not read English so she discarded the shirt. I did buy some clothes including a guayabara shirt which I now wear at certain parties. I only located my luggage when I returned to leave Panama. Braniff reimbursed me for the clothes.

At the base our civilian contact greeted us. Frank stated that the military consider this great duty and they try to extend their tour and/or return for another tour. Frank invited us to come to his home for dinner. Jack accepted but I preferred to have dinner in Panama City. At the hotel the front desk recommended a restaurant and called a cab for me. After dinner that cab was available outside the restaurant and returned me to the hotel.

In 1977, President Jimmy Carter sighed the Panama Canal Treaty promising to give control of the canal to the Panamanians before the year 2000.

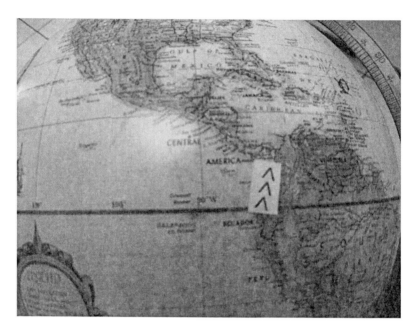

Figure 54. Panama.

Chapter 57

NIGHT VISION GOGGLES

"We own the night" is an axiom for the advantage our military night vision equipment provides. The Night Vision Laboratory of the Army Electronic Command, Fort Monmouth, NJ was the leader in the development of Night Vision goggles. I was in an auditorium for a night vision goggles demonstration. The lights were turned off and we could not see anything except the small red signs over the exit doors. When we put on the goggles that light was amplified so we could see people walking on the stage. (But everything was shades of green).

Years later I traveled to Fort Hunter Liggett, CA to evaluate the compatibility of the Night Vision goggles and my radar locator. The 101st Air Assault Division were flying training missions.

The nearest town is King City where I started my visit. The hills and scenery are beautiful and I told that to my contact. He said it was but after a couple of months you want something more. The Officers Club on the base was once the hunting lodge of William Randolph Hearst.

When briefing pilots I advise them that when my indicator detects an enemy radar to quickly descend to the nap-of-the Earth. During this training exercise the ground was covered with trees so some rotor blades were broken by hitting branches when descending.

During my visit I watched as the soldiers had an "aircraft down" exercise. They called First Sergeant Moats who told everyone to board a helicopter and they quickly flew to the crash site.

We first visited our son in Camp Ederle, Vicenza, Italy where Jay was in charge of a company. The First Sergeant there was Bridges and he was in charge of organizing the troops. Jay and Bridges remained friends even after Bridges retired.

One of the pilots gave me a sightseeing tour. We saw a golden eagle and the San Andreas Fault. It was a long wide ditch. The pilot saw a herd of wild cattle so he descended to ground level and chased the herd. When the bull turned to face us and put his head down we flew away. That area is remote and dangerous. Two soldiers were killed by wild boars during an over nigh camping trip.

When it was time to leave I told my contact that I planned to drive on a dirt road over the mountains. He said sometime there are hippies on the road and they may ask me for money. I hid my money except for a twenty dollar bill. At the top of one mountain I stopped and looked down to see the ocean. I saw a few whales and the scenery was beautiful.

Figure 55. Shield for Missile Alert LED.

Night Vision Goggles

Months later I was told that, "Red is Dead," in the cockpit because red lights are too bright if night vision goggles are used. My Missile Alert LED is red so I had a meeting with our machinists. I told them that I needed an add-on shield that I could slide over the LED. It must contain a pin hole so a small amount of light can be seen. They did it so I requested that they manufacture 200 more shields. Those shields were shipped to the Program Manager for Aircraft Survivability Equipment to be deployed to the pilots. I prepared a Modification Work Order to have the shields installed on all the Indicators.

Chapter 58

PEARL HARBOR

Jack from PM-ASE and I traveled to Hawaii to visit the 25th Infantry Division. We briefed pilots in Joint Base Hickham on Oahu Island. I gave a technical briefing that described the capability of the AN/APR-39 radar locator. Jack discussed his Pre Deployment Plan which was then approved.

On December 7, 1941 Japanese aircraft bombed and strafed Hickham to prevent our aircraft from following their aircraft back to the carriers.

We scheduled our meeting on a Friday with a return flight on Monday. We stayed in the Hale Koa, a twenty story hotel on Waikiki beach. Guests must be active duty or retire military or government employees with official travel orders.

I rented a surfboard at the beach. Then Jack and I drove to Diamond Head Mountain and we walked up the trail to the top. Then we started a trip around Oahu. We stopped at a small airport and we were the two passengers in a glider. The pilot landed the glider on the same area where we started. At the other side of the island the waves were enormous. Thankfully I never tried to surf there.

Figure 56. Joint Base Pearl Harbor-Hickham.

When our son Jay returned from Desert Storm we traveled to Australia. On our return we landed in Hawaii and I had booked a return trip to the USA for seven hours later. I rented a car and we had lunch in the Hale Koa hotel. Then we booked a boat trip to see the Battleship Arizona memorial.

Chapter 59

JAY WATERS' RETIREMENT

When Jay retired as the director of the U.S. Army Physical Disability Agency, his doctors gave him this skeleton. Jay drives a convertible. Sometime he puts the skeleton in the passenger seat and puts the top down.

Figure 57. COL Jay and his retirement gifts.

Chapter 60

MY RETIREMENT PLAQUE

When I retired I received this plaque "John Waters, Mr. AN/APR-39," and the photo.

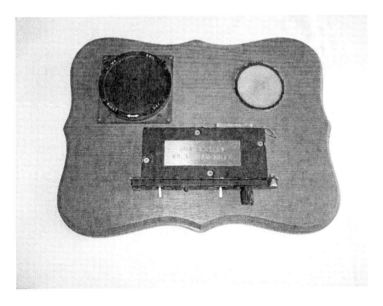

Figure 58.1. The circle contains the spiral winding of my antenna.

Figure 58.2. Another retirement gift.

Chapter 61

POST RETIREMENT

During my retirement I maintain contact with the AOC/Garden State Chapter personnel. Also the Info Age Center and Museum personnel which has EW publications and the history of Fort Monmouth. They also have an EW room with EW systems and photos.

My adult community has an active Veterans Club. We have a ceremony every Veterans Day and Memorial Day.

Bill would obtain a speaker from the Joint Base McGuire-Dix-Lakehurst. Their Public Information Center would provide active duty military personnel ranging from enlisted to Lieutenant Colonels. My son Jay was the only full colonel who was a speaker.

Bill was in the Battle of the Bulge and recently passed on. Now we also have a few civilians with military background who speak.

I am asked to recognize all the veterans in the audience at the ceremonies. George would show a video of the West Virginia Marching Band who plays the four Service songs. When the band rearranges into a tank, aircraft, etc. I ask those veterans to stand.

Joe, our club president, asked me to also recognize the spouses, children, and mothers of veterans. I told the audience that a line in a poem is, "They also serve who only stand and wait" That line also applies to mothers and wives of veterans. Then I ask all those women please rise and be recognized.

I also stated that our community has a woman who was on a ship during World War II. The captain was alerted that there was a German submarine in his area. He lowered the British flag then hoisted the American flag. At that time America was a neutral county and they were not attacked.

My brother sent me this statement which I read at our last Memorial Day ceremony. I received many positive comments about it, now I share it with you.

THERE IS A DIFFERENCE

Armed Forces Day is for those who currently wear the uniform.
Veterans Day is for those who use to wear the uniform.
Memorial Day is for those who never made it out of the uniform.

Figure 59. In front of the Clubhouse.

ADDENDUM 1 THROUGH 39

ADDENDA 1. CAMP EVANS

The Electronic Warfare laboratory was established in Camp Evans, Wall Township, NJ. Camp Evans was part of Fort Monmouth which is located about a distance of 10 miles to the North.

At Evans Colonel (Dr.) Blair was the director of the Signal Corps laboratory. They installed the U.S. Army first radar, SCR-268 on a hill in Highlands, New Jersey. It searched the shipping lanes for enemy bombers. (Later the SCR-270 detected Japan's aircraft about 40 minutes before the bombing started). Currently on that hill is the Twin Lights State historic Site which is open for tourists.

Evans became world famous in 1946 when the Diane Project modified SCR-271 radar transmitted a radio signal to the moon then received its return signal. Newspapers had front page headlines, e.g., "We touched the moon."

When I was in the Eight Grade I knew that when the moon was closest to Earth it was 239,000 miles away. I would calculated the different velocities that we would need to travel to the moon. (The Diane radar signal did it in 1.3 seconds).

The Diane project test site exists and is a tourist attraction. It is operated by the Info Age Science History Museum and Learning Center.

The center is open to the public and also accepts many student groups for overnight stays in their dormitories.

Fred who is the COO of Info Age accepted our invitation to speak to our adult community. He recommended that I give a presentation about Electronic Warfare. I am doing that for groups and college students. (That recommendation led to preparing this manuscript).

Evans also launched weather balloons. We watched as the balloons were set adrift then transmitted weather information to a microwave receiver. There was also a nuclear branch in a remote building in Evans. We looked carefully to see if Dr. H. was glowing when he attended our celebrations.

ADDENDA 2. SURVIVABILITY EQUIPMENT BRANCH

Circa 1960 Dr. Eugine Fubini was the defense policy person for the Cold War. As the Assistance Secretary of Defense he wanted to have military technical superiority when compared to the Soviet Union.

He funded the Survivability Equipment Branch of the Electronic Warfare Laboratory in the Army Electronics Command for the development of three systems. They were Countermeasures Set AN/ALQ-80, Countermeasures Set AN/ALQ-67, and Radar Signal Detecting Set AN/APR-24. He directed the Army Training and Doctrine Command to issue requirements for each system.

Six Engineering Development model of the AN/ALQ-80 were contracted and delivered. The models were installed on a pylon on the left wing of the Mohawk OV-1 aircraft in Viet Nam. They were designed to jam the 57 mm Anti-Aircraft Artillery Fire Can and the Fire Wheel radars. Reportedly one jammer jammed a radar directed AAA system. Then a contract was awarded for eighteen Limited Production models.

The AN/ALQ-67 was a Velocity Timed fuse jammer that was installed flush on the right wing of the Mohawk OV-1 aircraft. VT fuses were not used by the enemy so a production contract was not awarded.

The Radar Signal Detecting Set AN/APR-24 was competing with Air Force warning receivers and did not enter production. The Air Force AN/APR-25 was installed in Army fixed wing aircraft.

ADDENDA 3. ASA

The Air Force had the AN/APR-46 warning receiver which was replacing the AN/APR-25. The Army was scheduled to receive sixteen models. I called the Army Security Agency and asked if they would like to attend a meeting regarding the deployment of the models. The major told me they would attend the meeting and wanted all sixteen models for their fixed wing aircraft.

I was familiar with ASA and knew they had a "do it now and fix it later" mentality. The Soviet Union had a similar mentality. A Soviet General stated that the, "Best is the enemy of the good." The Army Delta Force also had a similar mentality. During my visit to their flight line in Fort Campbell, KY they had my radar locator on some aircraft and many other equipment on different aircraft. Captain Eddie may have been the only person who knew what was on each aircraft.

Conversely the Army Material Command wanted all tests performed and successfully completed. Our command was subordinate to AMC as was the Test and Evaluation Command. TECOM did their tests and reported all discrepancies. I wanted a report that also compared my systems to what was currently in the field. They did not do it.

The Program Manager, Aircraft Survivability Equipment is the PM in charge of electronic warfare systems and other survivability equipment. The PM stated that the Army will install the AN/APR-46 in fixed wing aircraft.

I visited Warner Robins Air Logistics Complex which is adjacent to Warner Robins, GA. I made initial arrangements to obtain the AN/APR-46. Returning to Fort Monmouth I call the person responsible for Interservice Logistics. He requested more information than I considered necessary but I gave him all that he requested. Then he called me and

148   *John Edward Waters*

stated that the Air Force would not provide any AN/APR-46 assistance. That made things more difficult for me but I contacted the Air Force and they agreed to add the Army to their AN/APR-46 program.

This was during the Viet Nam War and events were changing rapidly. New Intel information was received and changes were being made to AN/APR-46. After Change Number 12 the PM-ASE stated that our depot and other logistics functions could not respond to the changes. He canceled the program.

ADDENDA 4. GUARDRAIL AIRCRAFT

The OV-1 Mohawk aircraft and other Army fixed wing aircraft were being replaced by the Guardrail Common sensor aircraft. Guardrail is an airborne signal intelligence (SIGINT) aircraft. It instantaneously detects and locates microwave signals. That information is then transmitted to ground receivers of the battlefield commanders. Guardrail remains deployed and will continue operations until 2034.

Radar Signal Detecting Set AN/APR-39 was installed in the Guardrail. That made me a member of the Guardrail family and I was invited to their reunions. Herman was and is the Government "Mr. Guardrail" and is currently employed by the contractor. We remain in contact.

ADDENDA 5. SIGINT

Army Airborne SIGiNT was used when computers were first being manufactured. At that time there was only one computer at Fort Monmouth, NJ. It was the Burroughs B5000 that was introduced in 1961 and had remote terminals in Fort Monmouth. Project Engineer John was in the Electronic Warfare Laboratory and was writing code for his SIGINT system. He worked late at night so as to be the only person on a remote

Addendum 1 through 39 149

terminal. That reduced the likely hood of the computer aborting his program.

I did write programs on the EWL remote terminal near my office. My programs were stored on punch tapes. I had punch tapes named for each women who worked nearby. If one passed by I would ask her to type her name on the terminal keyboard. The computer screen would type, "I know you. You are ----. Do you know how to use this computer? She would type in "no" and my program would continue the discussion with her.

I enjoyed writing computer programs and software programing grew to become the Computer Software Center. I had the opportunity to work in the center but I enjoyed being a project engineer.

ADDENDA 6. LOWER MINIMUM DISCERNABLE SIGNAL

Herb was in my carpool and was working on a circuit that would increase the Minimum Discernable Signal of radiated signals a couple of decibels. Herb was named Fort Monmouth Engineer of the year. I recommended that he contact NASA who would be very interested in a circuit that would increase the MDS of their receivers. He said he would not do it.

Years later I attended a course on diodes and mixers at Georgia Tech University in Atlanta, GA. I contemplated what Herb's circuit could be. If it was used to detect a Continuous Wave signal it would easily prove that the circuit lowers the MDS and Herb would have received many more awards. The NASA signals have very long pulse widths (Which has very narrow bandwidth receivers to reduce noise) and it could also be successfully demonstrated. If it was demonstrated with narrow pulse widths there could be problems proving results.

I was impressed with the George Tech professors. They do consulting work so I kept them in mind if I had a problem in their area.

ADDENDA 7. EWL PERSONNEL

The Electronic Warfare Laboratory employed about one hundred engineers. A few had different personalities. Dave received an all expensive paid to attend MIT and obtain a Master's Degree. After receiving the degree he accepted a position with a military contractor.

Dave was a good salesman. He attended a Defense Advanced Research Projects Agency (DARPA) conference in Rosslyn, VA. He called me and told me I needed to send him a proposal for a delay line circuit. I did FAX him a proposal.

DARPA accepted the proposal and would provide funds. We would be working with a company that has a relationship with the University of Illinois. DARPA sent me a letter stating the Electronic Warfare Laboratory will need to send confirmation that EWL would accept the program. I gave the letter to the Division Chief for the required signature. He looked at the letter then put in his in basket. That basket contained other letters and folders. He did not sign the letter and a week later DARPA canceled the program.

Two young EWL engineers were given a four foot tall robot to write code instructions for the robot. Then they both accepted positions with the Computer Software Center.

Two EWL engineers refused to do any work. One had a marriage problem and was depressed. The other did not receive the promotion that he wanted.

One engineer told me he was a music composer and composed a familiar old song. Bill said he used an unusual pen name. He received royalty payments for the song. Then the real composer who had that pen name sued Bill and the court made Bill give all the royalty payments to the composer. Bill lost his security clearance.

Ron and his wife worked in EWL and planned a trip to paddle a kayak across the country. Both ran marathons and loved the outdoors. They received special permission to save two years of annual leave (One year was the maximum we could save). Then they researched what others did for this cross country trip. They started their trip in the Pacific Ocean at the mouth of the Columbia River in Oregon. They encounter an unexpected strong current and they would need to paddle against the current for hundreds of miles. That would require many extra days so they cancelled the voyage.

Mr. B was a Director of the Electronic Warfare Laboratory. The number of engineers and missions increased when he was director. He was the son of Colonel B. who was in charge of the development of Army radar. Some said Mr. B would doze at briefing. He never did that at my briefings.

Lou was the Project Engineer for the AN/APR-44. That radar warning system detected Continuous Wave radar signals from surface to air radars. The Egyptians deployed the Soviet SA-6 CW radar systems against the Israeli Air Force. The IAF had no CW radar warning systems on their aircraft which resulted in significant aircraft losses. Their After Action Reports resulted in changes in tactics and maneuvers.

Lou and I swam laps in the Monmouth University swimming pool during lunch time. I was the first swimmer to use the shower room. One day I entered the shower room and there were four black men using the showers. All were taller than six-foot seven. They were basketball players for the New York Knicks. If they started a conversation I was going to say that I was their new point guard.

The Knicks did their practice and preseason games at MU. I called MU for tickets to watch a game. I mentioned that I was a contributing alumnus and wanted three tickets. I received three tickets and passes to the locker room. After the game Jay and his friend Mark came with me to the locker room. The Knicks were in the locker room and their starting center asked Mark to get him a towel. He did and I am sure Mark told that story many times.

At a Bring your Child to Work Day a co-worker introduced me to her son. Tahj was six foot seven inches tall and the basketball star on the Red Bank Regional team. My son Jay also attended RBR and we both attended some of Thaj's games. In college Tahj was on the Marylyn University team that were the NCAA National champions.

152   *John Edward Waters*

Tahj is now the coach of a private high school basketball team that has two McDonald All Americans and is rated the Number Five team in the country. He and my son Colonel Jay were both inducted into the RBR Hall of Fame.

Ray our Division Chief had an anechoic chamber installed in our lab area. It was a small 9X7 foot room. When it was completed the contractor placed a dollar bill between the door and the door frame. Then the quiescent chamber had a detectable RF noise signal.

I did use the chamber to test my antennas. I placed a strip of LED stage lights between the transmit antenna and the receive antenna. When an antenna transmitted the lights illuminated. That looked impressive but I considered it unprofessional and did not use it. The lights were saved for Christmas parties.

ADDENDA 8. RADAR SET AN/PPS-5

Our EWL director requested a video about the AN/PPS-5 Radar Set be produced. He hired professional actors to perform at EWL. One actor was dressed in a soldier's uniform. He gave a battlefield assessment and the role that the radar set played during the battle. I had a brief silent part but did not see the other parts of the video. The director did not like the video and it was never distributed.

Bruce was a technician who worked for me on some of my projects. He was an amazing person although sometimes he was difficult to work with him.

He helped me when I purchased black gunpowder for a field test. Not sure how he knew so much about explosives. He also recommended that I purchase a book about transistors and their various applications. At that time I was more familiar with vacuum tubes and their various applications. I did have a book, "Introduction to Radar Systems" by Merrill Skolnik which I read many times.

Bruce caused two solid state devices to fail. I told him not to test our last device because I will contact the manufacture to come to EWL for a meeting. He did test the third device and it also failed. Our supervisor told me he almost fired Bruce for something he did during a field test.

Bruce's father was an engineer for Robert Moses who was known as the master builder for New York City, Long Island, Rockland and Westchester County. I remember Moses being in the news more than the mayor. Bruce told me Moses built parkways that had low bridges so trucks would not come into NYC from the northern counties. He also built highways in Long Island to promote urban development.

ADDENDA 9. NEW DEVICES

A colonel arrived at EWL for a demonstration of our equipment. I only remember what Tommy displayed. He had transceivers with antennas embedded in what looked like a rock and also an animal stool. Tommy displayed a device that could look through a door. The colonel asked if an X-ray device could be used. Tommy said no. That would require an X-ray plate at the other side of the door.

ADDENDA 10. NEW JAMMER

Many years ago there was a meeting at EWL to start an inboard radar jammer program. I was the youngest engineer in the group. They told me that they would all be retired before the jammer was deployed. I was currently the project engineer of the wing mounted AN/ALQ-80 jammer and this was an extra duty.

Then a younger engineer, X. L., was asked to be the project engineer for what would become the AN/ALQ-136 Radar Jammer. Eventually the AN/ALQ-136 entered production and was installed in Army fixed wing aircraft and rotary wing attack aircraft.

Like the AN/ALQ-80, the Air Force and the Navy had pod mounted radar jammers on the wing of their aircraft. Then DOD mandated a program for a single tri-service inboard radar. It became the Advance Self Protection Jammer, AN/ALQ-165. Development and production of the ASPJ was a joint venture between Northrup Grumman and ITT Avionics. (Northrup Hallicrafters produced the AN/ALQ-80).

At an ASPJ meeting there were forty persons in the auditorium. I was the only Army representative. The Army had its onboard jammer and resisted any replacement. I was mostly a spectator but did make one contribution.

There was onboard electrical interference on the aircraft. I recommended a blanking pulse to delete the interference. My AN/APR-39 generated a blanking pulse if there was interference. The blanking pulse was internal to the aircraft whereas the interference would propagate out to the antenna then travel along the transmission line. That was enough time to blank the leading edge of the interference.

ADDENDA 11. NELLIS AIR FORCE BASE

My assistance Project Engineer, Major M. and I had travel duty to Nellis Air Force Base, Nevada for a flight test with the AN/APR-39 installed in a UH-1 helicopter.

Nellis is adjacent to Las Vegas so we check into the Sands hotel in Las Vegas. Frank Sinatra was performing for his midnight show so we purchased tickets and attended.

The next morning we checked out then went into the gambling room to see if there were people gambling that early in the morning. Sinatra was gambling at a Black Jack table. He may have been drunk because he had a three and a five but did not asked for another card. A woman was near him and hoped to get one of his expensive chips or maybe get him. She would not get Frank because his wife Mia Farrow was seated in a chair against the wall watching him as well as his body guard.

At the Nellis AFB flight line the major and I boarded a helicopter and flew to a hill. The test radar started to track the helicopter. The pilot flew behind the hill and the radar broke lock. The pilot flew a series of coming out then in and each time the radar reacquired the aircraft. The reason for the test was to see how quickly the radar could lock onto the aircraft.

The radar locked on quickly. But I knew my radar detector would detect the radar before the radar detected the aircraft. Previously on a EWL test bench I connected a pulse generator to the input of one of the receivers. The radar detector immediately detected the signal. The detector will detect the first radar pulse although the strobe line on the Indicator would be dim. As the detector received more pulses the strobe line would become brighter. Within two seconds the strobe line was bright and the audio alert was heard.

From Nellis AFB we were scheduled to drive to Tonopah Test Range but decided to go through Death Valley National Park to get there. At a liquor store I purchased a six-pack of beer. I told the clerk we needed the beer to drive through Death Valley. He said I should buy two six-packs. I said we may not make it if we drank twelve cans of beer.

Death Valley is beautiful and surrounded by mountains. It is one of the hottest areas in the world and has the lowest elevation, 280 feet below sea level, in North America. At one time aircraft could fly lower below sea level than submarines. I left the car a saw a shallow spring with minnow size fish swimming around.

At Tonopah I asked pilots for comments about the AN/APR-25 that was on their aircraft. One pilot asked me if I would like to fly to the Area 51 facility. He could tell that I did not want to do it. He then said because I would need an oxygen mask and I had a beard it would be better not to go.

ADDENDA 12. AIRCRAFT MANUFACTURES

After my radar warning systems were installed in aircraft I would visit those contractors. I visited Sikorsky Aircraft Company, Stratford, CT. The warning system was installed in the Black Hawk UH-60 helicopter. I was concerned about the installation of the two tail antennas but the Aviation Command in St. Louis, MO is responsible for aircraft installations. If there was going to be any inaccuracy with target location then I would comment on that installation. One of the first production models UH-60 recently crashed in a nearby river so all flights were grounded. Currently more than 4,000 Black Hawk are in service.

I visited Beech Aircraft Corporation, in Wichita, KS. After landing I had lunch with the contractor. We had a beer and I said it lacked taste. He said it is 3.2% near beer and he put a shot of gin in the glass. Then it had lots of taste.

I inspected the RC-12 Guardrail SIGINT aircraft that had the AN/APR-39 installed. The tail section had been removed and there were two coax cables. They were cut off at different lengths so they could not be reconnected incorrectly when the tail was again assembled.

I visited Sikorsky Aircraft Company outside of Philadelphia, PA. My warning receiver was installed in the CH-47 Chinook heavy lift helicopter. Standing inside the aircraft I could smell motor oil. I was told it always has a motor oil smell. I decided not to do a flight test.

The CH-47 may be the biggest heavy lift helicopter in the World. But in a corner of the hanger there was a much larger helicopter. A ladder was required to step into the cockpit. That was the only helicopter of its type ever produced.

I visited Bell Helicopter in Fort Worth, TX. My warning receiver was installed in a UH-1 Heuy helicopter. I asked to go for a flight test. They called a pilot. He inspected the Huey and said a part needed to be tighten and he left. The part was tighten and the pilot returned then again inspected the aircraft and said he was going to go to lunch. I told my escort that my trip report will state that Bell Helicopter personnel do not assist visiting government personnel. He made a phone call and the pilot returned.

We flew around the area but I did not detected any radars. When the area was widen the warning receiver did locate a radar. I knew it was an airport radar because it only illuminated my indicator briefly every four seconds. (A ground air traffic control radar searches 360 degrees in azimuth every four seconds). The radar was at Love Airfield or Dallas Fort Worth airport.

I also flew a UH-1 in Germany. That aircraft was the first to have the AN/APR-39 wired and installed in Germany. Again I asked the pilot to fly

Addendum 1 through 39   157

around the area but we did not detect any radars. I asked the pilot to fly near a large microwave tower. I want to determine if could damage the receiver. Returning to the airport I listen on the intercom as the copilot stated there was a problem with the aircraft. The pilot asked if he should declare an emergency. The copilot said no and we landed without incident.

ADDENDA 13. FLIGHT LINE CHICKEN

The Hallicrafters Company delivered the first model of Countermeasures Set AN/ALQ-80 radar jammer to my office which I shared with three other engineers. They were surprised because they tell their contractors to maintain their systems. I was a new GS-13 Project Engineer and was going to do more "Hands On" with my project.

That radar jammer was transported to the Army Aviation Detachment in Lakehurst Naval Air Station. The detachment and their aircraft were in Hanger 5 which was once the hanger for blimps and the German zeppelin

The detachment was responsible for the cable wiring of a Mohawk OV-1 aircraft for the radar jammer. Once the wiring was completed one of their pilots would fly the aircraft to Eglin Air Force base in Florida.

I had been to Eglin and was assigned an office in the large King Hanger on the flight line. I ate lunch in a luncheonette in Fort Walton Beach. At my table there was a card. It stated that Jimmy Doolittle wrote his proposed bombing mission of the Japanese main island at this table. For my flight test a start date and a radar was assigned to track the aircraft.

The major in charge of the Lakehurst detachment told me that they cannot complete installation by the agreed date. I told him to keep working. Then Eglin called. They cannot meet the assigned date and proposed a new date. I said OK then told the Major that we just played, "Flight Line Chicken."

I contacted our Fort Monmouth photo branch and asked to have a photographer travel with me to Florida. O. C. was the assigned photographer. The radar was in a remote peninsula near Port St. Joe on the Florida pan handle and O.C. was a black man. I called a motel in Port St.

Joe and reserved two rooms. I mention that the other guest was a black man. The woman said that if a black man checked into her motel the other guests would leave. I called Eglin and they assigned one of their photographers to take pictures.

While in St. Joe there was an odor of burning wood. The odor came from a large paper mill located on the nearby bay. There was a lot of smoke coming from the mill. The population was about 3,000 persons and many worked at the mill. Our country was beginning to impose air pollution restrictions but they were not enforced in St. Joe.

ADDENDA 14. SECOND MOHAWK AIRCRAFT TEST

A fly-by flight test of the AN/ALQ-80 on the Mohawk aircraft was schedule against a radar in North East Florida. I would be the passenger in the co-pilot seat.

A flight ejection seat test and certification was required. So we flew from Lakehurst Naval Air Station in a twin engine propeller utility aircraft to a Navy airport in Maryland.

At the airport I was fasten onto a seat ejection test equipment. My instructions were to keep my elbows together then pull down the ring that was around my head when I was ready. There were about a dozen sailors who gathered to see what was going to happen.

I pulled the ring and was instantly about thirty foot above the ground attached to a shaft that was part of the equipment. I was shot up by a modified artillery shell. That shell is also under the two seats of the Mohawk aircraft. When that ring is pulled on the aircraft the person and the seat are propelled into the plastic roof of the aircraft and a parachute will open.

On our return flight the pilot stop his aircraft part way down the runway. He was not going to be airborne safely. He started the second takeoff with his tail wheel on the grass. We only needed to be a few feet above the runway because beyond the runway was only a beach and the Atlantic Ocean.

Addendum 1 through 39 159

The Florida flight test was scheduled and I needed to land my classified aircraft at an airport that would guard the aircraft. McCoy AFB is near Orlando. A person at the base stated that I could use the airport but it is closed and there is no JP-4 aviation fuel available.

I called Patrick Air Force Base and Cape Canaveral Air Force Station. They are home to the 45th Space Wing and the 920 Rescue Wing but they cannot guard classified aircraft.

I called MacDill Air Force Base in Tampa. They told me they would welcome me and they would park my classified aircraft on their flight line. It is guarded 24/7 by guards and dogs. If there is a problem with Cuba then their F-4 fighter aircraft would take immediate action.

I scheduled our first fly-by. The Mohawk was an unforgiving aircraft. Two had crashed within the last twelve months. There would be no copilot. Our flight plan is to fly across one edge of the Bermuda Triangle.

Frank my assistant project engineer said he would do both fly-bys. We both had young children so I told Frank that I would do the first fly-by and he could do the second fly-by. During the first fly-by the transmit light on the Control Panel of the jammer illuminated. That constituted a successful flight test. More detailed tests would be conducted at a future date. I told Frank that he did not need to do the second fly-by. He said he wanted to do it. (I knew he would want to do it).

ADDENDA 15. COMPUTER TESTS

The results from a radar jammer tests were sent to General Dynamics in Fort Worth, Texas. I visited that facility to observe the test results. There were large racks of seven foot high cabinets of computers in a cool area to reduce component failures.

The computer may have been the then popular IBM System 360 main frame computer. It weigh was 1,700 pounds and its memory size was 64 KB. My desktop computer has 4,000,000 KB RAM. The stock price for IBM shares were $600 for one share. So I purchased stock in Control Data Corporation/Cray which then had the fastest computer in the World.

Control Data Corporation/Cray used a multi parallel micro processing design computing system. Then other computer companies also did that and I lost money.

At GD I watched with the Information Tech as the data was downloaded. After 20 runs of the same event I stopped the test. The IT stated he needs 50 runs to confirm that the data is valid. I told him there was nothing to see and we did not need 30 more runs of nothing. I want you to start 50 runs where you first received information. I wrote the test plan and will sign whatever you want to make that change. He did make that change and the data revealed that Countermeasures Set AN/ALQ-80 did successfully jam the radar.

I stayed at a motel across the street from GD and it was adjacent to Grasim USAFB. The base supported heavy strategic bombing aircraft and I heard then fly overhead of the motel during the night.

ADDENDA 16. LINK AVIATION

I traveled to Binghamton, NY to the Link Aviation Devices, Inc. Link had flight simulators for pilot training and also refresher training six months later. I sat in the trainer and did a dive and a turn.

There was a proposal to install the AN/APR-39 Azimuth Indicator and Control unit in the flight simulator. I recommended that a program be installed that would display a radar strobe line of a radar beyond the Front Line of the Battle Area. Then if the pilot did not do a steep dive to the Nap of the Earth within three seconds he would fail the training.

If they installed a chaff dispenser the pilot was also required to drop chaff then make a turn away from the chaff cloud.

ADDENDA 17. CONSULTING ENGINEER

As a consulting engineer I had a few one day contracts. Once a manufacture called me and explained his problem. Then asked a few questions. I gave him the one answer that would solve his problem. I thought that he would no longer need me. But it did result in a contract and I traveled to his facility in South Bronx, NYC to speak to his engineers.

I was not sure if I could help them and was concerned that they would ask more questions than I had answers. However they knew so little that I spent most of the time at the meeting educating them about their project.

I applied for a consulting job at a local contractor's facility. Bob was the owner and asked me what I knew about RF power amplifiers. I put a block diagram of a radar system on the chalk board then discussed the three RF amplifiers. Bob said he has interviewed a few digital engineers but I was the analog engineer that he needed. I was listed as a staff engineer but was the acting general manager who could fire anyone when Bob was not available.

Bob's background was as an electronics technician but he had the knowledge to design and build small RF power amplifiers. It was a niche market and he had many orders.

He had an order for a large high power RF amplifier which was the size of a hot water heater. It would be delivered to the Fermi lab Particle Physics and Accelerator laboratory in Batavia, IL. I was told that this would be part of a project that could result in the director receiving a Noble Prize.

Bob's company was sold to another company in Lexington KY. I was again hired to work at EWL. I continued to work part time at my job and also part time at EWL. The pay was less at the job but they needed me so I worked part time for two extra months.

The new company did an inventory audit. They counted the nuts and bolts in their trays. After the transfer contact was signed they found a five foot high can of toxic material. I do not know how that was resolved.

One of the draftsmen told me he could easily get another job but months later when he was about to lose his job he told me he cannot find another job. I told him I saw a notice in the store window in the nearby town of Red Bank that they were hiring draftsmen. He stated that he knew about that company and their pay was very low.

New Jersey human resources personnel arrived at the plant to interview employees and help them find jobs. One secretary was happy because New Jersey was going to pay for her to go to college.

The new company offered to transfer technical personnel at their current salary. I was not interested as I was working part time with the company and part time with EWL.

Our Government has contracts for Small Business Set Asides. Only small businesses can bid on those contracts. A contactor tasked me to help prepare a proposal for a subsystem of a radar signal locator. I did help but we did not get the contact award. That radar signal locator subsystem was once a small business set aside but the Government did not receive any proposals. If I was tasked at that time the contactor would have received the contact award.

EWL awarded contractor support contacts for two years of support. Then they advertised a Requests For Proposal for a contract. The Engineering Professional Service Company won the contract then hired me to continue to support EWL. EPS also asked me to help prepare proposals. One proposal was to install a system in the High Mobility Multi-Purpose Vehicle. We needed to take measurements on the HMMPV (HMMPV/Hummer). I could not locate any HMMPV in Fort Monmouth but did see a couple when I traveled to the Garden State Parkway. Along the route there was an Armed Forces facility which had military vehicles in their parking lot. The facility was being used by a company of US Marines.

Frank was the President of EPS and we both visited the Marines. I told the duty officer that Frank was a retired Navy Captain and commanded an aircraft carrier. He allowed us to take measurements of a HMMPV.

I did separate billing when working on proposals. I also did not bill when I could not sleep thinking about my projects and writing down what I was thinking. I did not bill when thinking about problems while jogging on roads. After fifteen minutes I get a Runners High from endorphins. I felt good and was at my best for solving problems.

Addendum 1 through 39   163

Note that EPS was one of two bidders for the contract. The other company's copier malfunctioned and they were more than an hour late to submit their bid. They were disqualified.

ADDENDA 18. HORN ANTENNA

I was designing a circuit that was connected to a horn antenna. I needed to determine the gain of the antenna. There is an equation to determine the gain of a horn antenna. I entered the frequency band, the height and width of the antenna in the equation and obtained the gain.

Then I asked the head of the antenna branch at Fort Monmouth. He told me to use the equation. Then I asked John who was the head of the Calibration Lab. He looked at the antenna and told me the gain. John also stated that there was a mismatch between the antenna and the adapter component connect to the antenna.

Then I contacted a manufacture who builds modules that are used to open and close doors automatically. He mailed me one module with its horn antenna attached at no charge.

I connected the module to an LED lamp. That device was pointed at the door in a trailer that was my office. If anyone entered the trailer the lamp would be illuminated. Eventually I gave the module to my supervisor for her enjoyment.

ADDENDA 19. COCKPIT INTEGRATION

I was a member of a group to recommend the integration of all cockpit displays on Army helicopters. The AN/APR-39 Radar Signal Detecting Set Control and Indicator subsystems would be integrated. I attended a meeting in Fort Eustis, VA which was conducted by the Army Aviation Applied Technology Directorate under the Research, Development and

Engineering Command. The UH-60 Black Hawk, AH-64 Apache, and OH-6 Cayuse were candidate aircraft for the cockpit integration.

Years earlier I visited Hughes Aircraft, the manufacturer of the OH-6 aircraft. At that time they were located in Playa Vista, CA. My escort told me that the hanger that we were in was where the Spruce Goose aircraft was built. Spare parts were stored there. The Goose was the largest flying boat ever built.

At that time the Spruce Goose was displayed on a pier in Long Beach, CA. The Queen Mary was docked at an adjacent pier and was a hotel. I stayed overnight on the Queen Mary. If you had Government Travel Orders you were given a large deduction for a cabin. Currently that aircraft is located in the Evergreen Aviation and Museum in McMinnville, Oregon.

The Hughes Aircraft Company faculty assets were transferred to Tucson, Arizona. The Arizona facility was where we had our next integration meeting. The meeting was scheduled for a Monday so Barbara accompanied me. We left Newark airport on a Friday and I checked into a hotel in Phoenix. When the desk clerk noticed that I lived in New Jersey he remarked that the New Jersey weather is uncomfortable because of the high humidity. Outside the hotel the temperature was 105 degrees but I did not mentioned that to the clerk.

The next morning we planned to visit the Grand Canyon. On a mountain near Flagstaff it started to snow. As I descended the mountain there was a head-on collision between two cars at the bottom of the mountain. I started to decelerate slowly so as not to skid. However the car behind us was skidding and out of control. I would not go any faster because I would crash into the cars below. Soon the car hit us and pushed us into the road railing. The driver jumped out of his car then told us that he was an FBI agent. Then he ran up to the collided cars.

After a policeman arrived he asked me what happened. I explained and the agent stated that it was his fault. Both cars were drivable so I followed him to a National Car Rental agency. They gave me another car but we cancelled our trip to the Grand Canyon.

When our son Jay was with an airborne brigade in Italy he was at a party and mentioned that he had never been to the Grand Canyon. All the

Italians stated that they had been there including Anna who would be his future wife.

The next day we had dinner in the Princess Hotel in Scottsdale. We knew the dinner would be expensive but it was the best service we ever had.

On Monday the meeting was at the Hughes Aircraft Company in Tucson. That evening we drove twenty miles into the desert to a restaurant that caters parties. It was beautiful scenery and there was a nearby mountain that people climb.

Barbara and I were the first people to leave. There had been an isolated rain shower which caused a ten foot wide pool of water on the road. I was not going to cross that waterway. I read that two soldiers drowned trying to cross a waterway. Then a car arrived and did cross the waterway. I followed the car.

ADDENDA 20. ARMED SERVICE ELECTRO STANDARDS AGENCY

My first business trip was while working for the Armed Service Electro Standards Agency. I flew from Denver over the Rocky Mountains on a two engine propeller aircraft with Frontier Airlines. At Grand Rapids I drove to a contactors plant to inspect his facilities.

He was fabricating voltage meters per a military voltage meter specification. I was mostly interested to see documentation that his voltage test equipment was calibrated to the standard voltage at the National Bureau of Standard, now the National institute of Standards and Technology (NIST). The equipment was calibrated and ASESA would send him a certification.

I stayed in a small motel. When retiring for the evening I noted that the door could not be locked. I tilted a chair then leaned it against the door. In the middle of the night the night clerk tried to enter the room and knocked the chair onto the floor. He said he was sorry. In the mourning the clerk said he thought the room was unoccupied.

Years later I visited NIST in Gaithersburg, MD. I inspected a Soviet radar directed weapon system that was being calibrated and tested. I climbed up to the roof of the system while it was radiating. I often wondered if the RF antenna back lobe radiation did something to me. Regardless it was a dumb thing to do.

ADDENDA 21. TEST EQUIPMENT

Frank was my second assistant Project Engineer and he would take actions without any discussion with me. I found a secret document in my safe that had been ripped almost in half. I also noticed that there was a second secret document that super seeded the ripped document. At first I was shocked but after further thought I realized that it was a good idea because I would never use that ripped superseded document.

I received a call from the R&D command inventory manager who wanted to know where six Hewlett Packard signal generators were located. I told him I did not know anything about those signal generators but would look around. Months later I was given a letter addressed to our Electronic Warfare laboratory director that wanted to know the location of the generators.

I asked Frank and located five of the generators in an unoccupied building on the Post then I sent a reply that was signed by our EW director. I stated that Hewlett Packard never made one model listed in the letter and provided the model number that they were interested in finding. I stated that all the generators are considered obsolete and the National Guard did not want them. Frank shipped the missing generator overseas to a combat unit. That letter ended the much ado about nothing.

There was a Fort Monmouth surplus office that was in operation for many years. That may be where Frank picked up the generators. I purchased an Eico oscilloscope there and brought it to Barbara's home to show her mother. I attached a headset to the input. I told her mother that if

Addendum 1 through 39   167

she told that black disc her name she could observe her name on the screen. She did and a green complex waveform was on the screen. The next day Barbara told me that her mother thinks John is an unusual boyfriend.

Years later the secretaries were using electric typewriters and turned in their manual typewriters. I purchased a Remington and a Royal manual typewriter from the surplus office. I used one for my typing and gave our children the other one to bang on the keys.

The Department of the Army initiated a program to have all personnel turn in all unused items in their possession. Some of us called it the, "Save a dollar and waste millions." I let the Avionics Laboratory borrow an AN/APR-39 unit. They turned it in. I did not realized that until an engineer in EWL told me he purchased it and sold it to a contractor.

The EWL deputy director inspected our area. He wanted to know why we had flashlights in the safe but the confidential AN/ALQ-80 jammer was on a test bench. I told him we are going to modify the flashlights then use them for flight line testers. We did not want anyone to steal them.

ADDENDA 22. GPS AND THE MARINE CORPS

I had a project that needed to transmit signals from multiple identical systems simultaneously. The transmitted signals needed to be precisely timed and synchronized with each other. We used the Global Positing System to do it which in addition to providing locations also transmits time from their atomic clock. The Department of Defense was in charge of the GPS program. The Fort Monmouth Electronics Command worked with the USAF office in California. Our GPS Office was concerned that many new Avionics systems would add a GPS module to their system. I told the GPS manager that I was only interested in locating enemy radars systems.

I did attend an all-day GPS conference at the Army Missile Command in Redstone Arsenal, AL. One speaker stated that if the GPS receiver receives less than four signals it could result in a transmission error of the true location. The location could be a large error such as a location in a different country. The user should turn off the receiver and try again until receiving the correct signal. In mathematics that is called the Axiom of Elimination. Currently there are thirty three GPS satellites in orbit. At that time there were five satellites.

Months later Ray our Chief Engineer stopped in my office. He received a call from a naval admiral that he worked with on programs. There were Marines in Haiti stationed as peacekeepers but when their vehicles drove under a bridge their GPS malfunctioned. I told Ray to tell the Marines to turn the GPS receiver OFF then turn it ON again. That is now called, "reboot."

Note that because we have thirty-three GPS satellites an enemy attack against one satellite would have only a minor disruption of GPS operations.

ADDENDA 23. TEST PLAN

Jansky and Bailey, Inc. prepared a test plan for a flight test of the Countermeasures Set AN/ALQ-80. Norman from J & B was the author and also prepared the test report. Years later I submitted a Request for Proposal for another flight test. I requested that the RFP should be a sole source to J & B and Norm should be the lead engineer.

The contracting officer called me and stated that we cannot name an individual on a contract. He changed it to "Norm or equivalent." Norm was no longer working for J & B and I considered their test plan to be inadequate. I rewrote it. The J & B supervisor told me I should list myself as the author. (I did not do it).

ADDENDA 24. NAVY MAINTENANCE PROGRAM

Two inspectors from the Congressional Government Accounting Office were at my desk (See Chapter 14a) when my phone rang. It was a representative from the Navy calling about starting a joint maintenance program. I should meet Tom in the Comfort Inn near the Minneapolis airport. He would mail me a map. I was concerned about the GAO investigation so I quickly hung up.

Tom did mail me a map but it was mostly illegible. I did schedule a flight to meet him. At the airport I rented a car and drove to a Comfort Inn. They did not have any reservation for me or Tom. After looking again at the map it was for Indianapolis not a map for Minneapolis.

I called the airport and booked a flight that was leaving in 30 minutes then drove back to the rental company. It was closed for lunch so I dropped the keys in a drop box. Then looked for a ride to the terminal. It was very cold and there was a car nearby that was warming up. I tapped on the driver's window and told him I would give him five dollars to drive me to the terminal. He did drive me but would not take any money.

I made the flight but my luggage did not. The next morning my luggage had not arrived. I left a note stating that I would return. I left my souvenir spoon from Minneapolis on the sink counter. When I returned the spoon was gone but my luggage did arrive. Currently I have more than a hundred souvenir spoons but none from Minneapolis.

I inspected the Navy's maintenance facility. It was impressive but we never did have program with them.

I did have chance to go to the Indianapolis 500 stadium and made one lap around the track. They have a minibus that drives tourists around the track. I still have that printed business card stating that I completed one lap.

ADDENDA 25. SORTIE FLIGHT TEST

Arrangements were made to fly a sortie from the Grumman Aircraft Engineering Company airport in Bethpage, NY against a radar at Stewart Air Force Base in Newburg, NY. The aircraft had Countermeasures Set AN/ALQ-80 under a wing. I invited Barbara and our three-year old daughter to stay with me for the short drive to Stewart AFB.

170 *John Edward Waters*

During the test I was in the radar trailer and watched the indicators. The pilots did a couple of dry runs with the jammer turned OFF. Then I asked them to turn ON the jammer and repeat the runs. The A-scope was no longer giving the correct range and the PPI-scope was indicating breaks locks and could no longer give the correct azimuth location.

That evening we went to an up-scale restaurant in Newburg to celebrate. There was live music but I usually do not dance. This time I did dance as Nancy watched me dance with her mother.

Months later there was an Electronic Warfare conference at the Grumman facility in Bethpage. We inspected the EA-6 Prowler stand-off jammer used in Vietnam. When the Russians introduced the SA-2 Fan Song missile guidance anti-aircraft we responded with our Wild Weasel program. The aircraft were the F-4 Phantom and the F-105 Thunder Chief aircraft. Their tactics were to allow one aircraft to be tracked while another aircraft would locate that radar and fire an anti-radiation guided missile. The decoy aircraft had the AN/APR-25 warning receiver onboard. When the missile was launched the SA-2 missile guidance radar was turn ON. The waring receiver detected the missile guidance radar signal and a warning light was illuminated. Also a distinctive audio tone was heard on the head set. The pilot would then take an evasive action. My AN/APR-39 has the same feature and I told pilots it was their last chance to take an evasive action.

The F-105 fighter-bomber had a high casualty rate during the Vietnam War. The aircraft flew 20,000 sorties. I was at a meeting with an F-105 pilot. He had a 300 mission patch on his uniform. He told me that an optimist is an F-105 pilot who gave up smoking for health reasons.

EWL developed Countermeasure Set AN/ALQ-80 (V-1) to jam radar guidance anti-artillery aircraft systems. The Hallicrafters Company developed the AN/ALQ-80 (V-2).

The (V-2) added a subsystem to the jammer which extended the length of the jammer. The (V-2) would jam that SA-2 missile guidance radar. The V-2 program was given to Sam to be the Project Engineer. He was a very no-risk person and told me that if there was any problems it was my fault. I told him OK. The program remained in the development phase.

ADDENDA 26. ALL FLIGHT TESTS

The flight tests at Eglin AFB, FL were the only fight tests that I had a contractor prepare a flight test document and a test report. The other test sites I prepared the flight test document and a trip report.

There were several occasions when I scheduled flight tests at Patuxent River Naval Air Station, MD. We used their Integrated Battlespace Simulation and Test facility.

Pax River is the HQ, Naval Air System Command and an operational base for the P-3 Orion aircraft. The P-3 is a four-engine turboprop anti-submarine aircraft. The P-3 tail boon is used for magnetic detection of submarines. It has a sonobuoy sonar system that is dropped from the aircraft.

I did not see any P-3 aircraft because they flew over the Atlantic Ocean whereas we flew over the Chesapeake Bay. A fighter aircraft did fly under our helicopter. We were later told that the pilot declared an emergency and was heading towards a runway.

My test plan allowed the helicopter to perform various tactics to avoid the radar and perform other tactics to break lock after the radar acquired the helicopter. I asked the pilot to hide the helicopter behind a lighthouse. He did and we descended until water was sprayed on the windshield. We did break lock.

We stayed in the adjacent town of Lexington Park. I would always eat breakfast in the Brass Ass restaurant. Once I brought my family for a flight test and we ate breakfast in the Brass Ass. The kids enjoyed saying that name without getting in trouble.

ADDENDA 27. QUICK REACTION CAPABILITY

During the Vietnam War some actions were completed quickly. I was given a message from our higher command, the Army Material Command. They needed a contract award for an Electronic Warfare system. AMC cited this QRC as a Number 1 priority. Funding would be provided at a later date. I never had or even knew of any Number 1 priority.

I asked a coworker. He said if I saw something I could take it and I could do whatever it takes to complete the task quickly. I prepared a Request for Bid and hand carried it through channels for signatures. One signee stated he would need time to review the request. I told him I wrote down the current time and would write the time he sighed off then attached it to the RFB. He looked at it then signed. There was a quick contract award.

Then were received a message from AMC. There were no funds available and we should use funds from our current assets.

The Hallicrafters Company had a blanket QRC contract with the Air Force for anything related to Electronic Warfare. I did not request anything but the Air Force did request radar jammer systems and modifications to their systems as new Intel information was received.

Hallicrafters and other military contractors had government inspectors on-site. I would meet with the inspectors prior to my meetings with the contractors. The inspectors had a stamp which was used as approval for modules and items that were ready for shipping.

Sanders Associated had a similar QRC program with the Navy. Navy personnel were stationed in the Sanders facility and could sign contracts. They tried to prevent me from coming into the building. Additional information had to be FAXED to their office before I was allowed in the building.

A fellow project engineer from EWL was at Sanders and invited me to be an observer at his field test. The test was in an open field and there was a dense tree line about 100 yards away. Behind the trees there was a replica of an enemy soldier looking at us with his binoculars. Then a laser type system was turned on. The test results were that if that was an enemy soldier then his eyes would have been impaired or totally blinded if he continued to look into the binoculars.

The test was a success but higher command cancelled the project. They considered it inhumane to blind an enemy. My son COL Jay stated that we

Addendum 1 through 39   173

cannot use tear gas against the enemy. Yet it is known that to wound an enemy is more beneficial than killing an enemy.

Routine contract awards were finalize with a meeting among the contactor, the Contracting Officer and the engineers. At a meeting we would discuss the line items in the contractor's proposal. The CO chaired the meeting. I would ask the CO to begin with certain line items. I studied those items and could rebut the contractor's man-hours for those items. Once the Moisture Fungus Proof report was delivered to me before the negotiations. It was a two page report but there was an excessive number of man-hours charged to that item. I asked the CO to start the negotiations with that line item.

At one negotiations there was an impasse. I asked Max our Division Chief to get involved. Max had a meeting with the contractor's manager who was not part of the negotiations. Max asked the manager to make an offer. The manager did and I accepted the better offer.

ADDENDA 28. INFRA-RED COUNTERMEASURES

The first combat use of the enemy SA-7 Strela infra-red missiles were in 1969 by Egyptian ground troops against the Israeli Air Force. The kill ratio was low against their fast combat jets. Even hits often resulted in the aircraft returning safely to the airfield.

Starting in 1972 in Vietnam the SA-7 kill ratio was high against helicopters and propeller driven aircraft. Some areas were declared no-fly zones to reduce losses.

Use of flares was effective but only for short durations due to the limited number of flares on board (See Chapter 25). PM-ASE had the helicopters modified to change the exhaust duct. The duct originally directed hot air to flow out to the rear of the aircraft. The missile locked on to that heat. The duct was bent 90 degrees to allow the hot exhaust to flow up to the rotor blades and be widely dispersed. Thus there was no hot spot for the missile to lock on.

In addition the Electronic Warfare Laboratory awarded contracts for infra-red jammers. The Infra-red countermeasures jammers transmitted an IR Pulse Radiated Frequency that was slightly different than the known PRF rotation of the missile. That error resulted in the missile being off target. Unlike flares the IRCM jammers operated continuously throughout the mission. Contracts were awarded to what would become nomenclature AN/ALQ-144, AN/ALQ-147, and AN/ALQ-157.

The AN/ALQ-147 is an airborne Countermeasures Set to counter infrared seeking missiles. It is produced by Sanders Associates. The AN/ALQ-147 was mounted on a wing pylon in a drop tank.

The AN/ALQ-157 is produced by Loral has two jammers one on each side of larger aircraft.

Note that the ANALQ-156 is a pulse Doppler radar system. It will automatically command the M-130 dispenser to eject its Infrared flares.

The AN/AAR-47 Missile Warning System produced by Loral/BAE Systems is a passive anti-missile system. The system detects the Ultraviolet output of a missile then initiates both a visual and audio warning. It also activates the on-board flare dispenser. More than 5,000 system were manufactured.

The AN/AAR-47B (V2) added the ability to detect incoming rocket propelled grenades.

Rich was the Project Engineer for the AN/ALQ-144 Infra-red Countermeasures Set. It would jam the SA-7 Grail. (Then later the AN/ALQ-144A would jam the SA-14 Gremlin, the SA-16 Gimlet and the SA-18 Grouse) The AN/ALQ-144 was deployed during the Gulf War. All were man portable. The AN/ALQ-144 was the size of beer keg and was often called a Disco Ball. The AN/ALQ-144 was usually mounted near the exhaust duct of the aircraft. Rich told me that the first system was delivered to the highest priority helicopter. To me that meant the presidential helicopter. That helicopter has many countermeasures onboard. There is a box over the rotor blades of that aircraft. It looks like the box I had installed over a helicopter as an option if a complaint was verified (See Chapter 49).

Addendum 1 through 39 175

I was contracted to work for Rich and often sat at his desk. He would often be on the phone with military personnel about all aspects of the AN/ALQ-144. That was a surprise to me because I had two thousand AN/APR-39 systems deployed and fewer phone calls. At presentations I even distributed a brochure about the AN/APR-39 which contained my address and telephone number.

Sanders Associates manufactured more than 8,000 AN/ALQ-144 IRCM systems.

The AN/ALQ-157 from Loral/BAE had a higher output power and a subsystem was mounted on the two sides of larger aircraft, e.g., CH-47 Chinooks.

We also have man-portable air defense missiles. The most notable is the FIM-92 Stinger missile. The Stinger is manufactured by Raytheon missile Systems. The Stinger has a range of more than 15,000 feet and its Infrared seeker has a superior anti-jamming circuit.

During the Soviet War in Afghanistan the CIA provided Stingers to the Mujahedeen fighters. The deployment was resisted by the CIA for various reasons. But based on urging by Congressman Charlie Wilson, President Reagan authorized the shipment of Stingers to the Mujahedeen fighters. This resulted in more than 260 kills of fixed wing and rotary wing aircraft.

The American Wings Air Museum in Blaine, Minnesota had on display IR systems and many other Electronic Warfare systems. The museum is no longer operating and their assets were transferred to other organizations.

ADDENDA 29. NEED FOR AN IRCM PROJECT ENGINEER

EWL personnel asked a contractor, Modern Technologies Inc. to hire an Infra-Red Countermeasures engineer. I was their consulting engineer so Larry asked me to find someone.

The engineer would work on one of the EWL IR programs. There are not many IR engineers in the Army Electronics Command. Bob was an IR

176   *John Edward Waters*

engineer who worked in EWL and recently retired. I knew Bob and planned to contact him to see if he would work for Modern Technologies.

I once visited Bob at his desk and he showed me how he would chart a company's performance then he might buy or sell their stock. He had books with graphs of stock prices that went up or down. If a stock went up for a certain period of time he would buy the stock. I did notice that he spent a lot of time doing stock trading.

(Years later there were day traders who were making a lot of money. Then the market fell and they lost their money).

I invited Bob to be my guest at an Old Crows luncheon. He said that he would like to join Modern Technologies. Larry spoke to the EWL people. They did not want Bob.

ADDENDA 30. RADIATION

The Electronic Warfare laboratory had an early requirement to develop sensors for Remotely Pilot Vehicles (now Unmanned Aerial Vehicles). I once stopped in the EWL Commodity Office. On the wall there were two large impressive pictures of the USD-5? and the USD-6? UAVs. Both were the size of a refrigerator. The USD-6 was jet propelled. Both were never developed.

The Army did developed the Lockheed MQM-105 Aquila, a seven foot UAV which was launched off a truck. The payload was a TV image and laser designator. A friend told me that there were major reliability problems with the UAV. The Army requested a production contract but the Department of Defense cancelled the program.

I was first involved with UAVs when someone from the Department of the Army called me. He wanted to know if EWL had a one-pound or less sensor that could be the payload for a UAV. I knew that our Radiation Division had a light weight dosimeter device that could be carried on a person to record the quantity of ionizing radiation, e.g., gamma rays that a person had been exposed and accumulated over a period of time.

Addendum 1 through 39 177

I met with Ed the Radiation Division Chief who said they could provide a sensor. After installed in a UAV the sensor would record the amount of gamma rays over a designated area. I told that information to the caller and told him that Ed would be his point of contact.

I use a taxi service and Cowboy is often the driver. He worked in the Oyster Creek Nuclear Generation Power station in Lacy Township, NJ. There he wore a dosimeter photographic film badge on the outside of his clothing. If any dosimeter exceeds a certain level of radiation that person would be medically treated to repair or replace damaged cells.

My son Jay wore a film badge during his deployment in Desert Storm. (More than 10,000 personal dosimeters were issued during Desert Storm). Jay was more concerned with a chemical attack and was taking a daily pill to minimize any chemical effects. He was aware of the enemy Scuds that flew overhead and exploded a short distance away.

Jay did say that he does have an effect after living in the desert for months. When he wakes up in the morning there are small grains of sand near the corner of his eyes.

During World War II on 6 August 1945 we dropped an atomic bomb on Hiroshima. Today Hiroshima has a population of more than one million people and is a modern city with tall buildings and wide streets. The current radiation level is very low which is about the same level as most of the world.

ADDENDA 31. ENGINEERING COMMITTEE CHAIRMAN

One of my home smoke alarms was beeping so I changed the batteries in all the smoke alarms. The mounting bracket of one smoke alarm was detached from the ceiling and the two electrical wires were separated from the smoke alarm. I opened the Circuit Breaker panel and turned OFF the CB for that circuit.

I am also familiar with all the CB panels in our Lake Ridge Homeowners Association Clubhouse and the Annex. As Chairman of the Engineering Committee I checked the hundreds of CBs in the Clubhouse and Annex. Then I taped a full page Power Point list on the back cover of the door panel which identified each CB. After I found a CB circuit location, Freddie would tape an identifying label on the switch or receptacle.

Once a switch in the Clubhouse was smoking and was too hot to touch. Freddie was in the Clubhouse and looked at the switch label then turn the associated CB to OFF.

The committee prepared a two hundred page maintenance manual. I submitted about half the pages. That task was helpful in writing this manuscript.

ADDENDA 32. ELECTRONIC BORDER FENCE

President Trump has proposed a see-through metal wall along the Mexican border. That type of wall allows the use of the REMBASS passive systems and my radar system to detect intruders.

I designed a low cost radar system to detect personnel. It will detect and locate any movement across the border near the fence. The system will include the motion sensor used to automatically open and close doors. It is a low cost module. The manufacturer mailed me one module at no cost. I connected the module to an LED bulb and placed it on my supervisor's desk. She could see when anyone entered her room.

Engineering Professional Services Inc. received a contract to support the Remote Battlefield Sensor System, AN/GSQ-187 (REMBASS). I was aware of the program and discussed it with Frank the president of EPS. I also discussed it with the EWL project engineer. (I gave him a copper wrist bracelet for the arthritis in his hand). REMBASS has passive sensors to detect targets using seismic/acoustic/thermal and magnetic changes. It was used as part of the McNamara fence line at the Demilitarized Zone that spanned all of North Vietnam then into parts of Laos.

ADDENDA 33. MY PROPOSAL FOR A BARRIER ON THE BORDER

To: President Donald Trump
General Patrick Shanahan. Acting Secretary of the Army
Mr. Brian Kolfage, President, WeBuildtheWall
SLSCO Ltd.

The completed manuscript, "Aircraft Electronic Warfare Systems from a Project Engineer" by John Edward Waters will be sent to a publisher. This is Addenda 33.

There will be problems building a contiguous wall across the Mexican border. There are four National Wildlife Refuges containing threaten and endangered species, and at other areas it will not be cost effective to build a wall, also land owners will want access to the Rio Grande River for their family and livestock. So there will be areas without a wall. This is a proposal to provide a barrier for those areas.

The barrier will consist of PVC ridged pipes. They will be spaced across those areas to be wide enough for current Border Patrol ATVs, wildlife and domestic animals to ingress and egress. Wider vehicles will not be able to pass between the pipes. Intruders from the Mexican side can pass but the closest of the nine Border Patrol stations will be quickly notified of their presence.

Many pipes will be 1.5 inch in diameter with two five foot length pieces glued together. One end will be in a coffee can filled with cement and placed three feet in the ground. The other end of some of the pipes will have a housing fasten to a spigot. Wider pipes will be used if there is a likelihood of vandals trying to damage the pipes.

The housing will contain two motion sensors similar to the sensor used to open doors except they will have more output power to obtain a wider field of views and the antennas will be oriented to also have a wider field of view. I designed and built a motion sensor circuit and its output illuminated an LED bulb. An LED bulb will be used for the field tests. The antennas will be oriented to point at 45 and 135 degrees. (My AN/APR-39 radar detector antennas are also pointed 45 and 135 degrees on the nose of aircraft). The output of both sensors will be combined using a TEE connector or terminal board.

The output from the TEE connector will input the Iridium Extreme 9575 Satellite phone or the equivalent. The output from the phone connects to its optional antenna which is fasten to the top of the housing. That email signal will be received at the nearest Border Patrol station and will contain the location ID number of that pipe. Also on the top of the housing is the solar cell charging panel for the lithium ion battery pack.

Sincerely.

The above letter is brief because many subjects should be discussed during meetings. Land owners may elect to have a narrow opening along the wall and only one pipe with a motion detector sensor. The owner could also be notified by email if there is an intruder. Also the owner should be offered a commercial radar that would be pointed at the opening in the fence. There is a manufacturer in New York who may manufacture a radar that could meet there needs. The owner should also be offered a pair of binoculars. If these things are offered then the owner may agree to allow an easement. There would be no need to obtain an eminent domain decree.

PVC pipes are sold with a choice of two colors. A gray color pipe is less observable to the intruder. A white color is more observable to our vehicles driving on the border.

The enclosure on the pipe is an aluminum alloy which is sold in many sizes in length, width and height. We would also want a flange to attach it to the spigot. The high ambient temperature requires special consideration. The enclosure should be painted white to reflect the heat. Holes should be drilled in the sides and also into the rear to reduce the heat inside the enclosure.

There will be openings for the camera and the antenna. Both the camera and the motion detector module will need to be mounted on wedges to observe the ground/

Construction should consider methods to prevent birds from nesting in the enclosure and also to prevent insects from using the enclosure.

The camera could be operated continuously or only when the motion detector turns it ON. Both will required a person to retrieve the data.

ADDENDA 34. RONALD REAGAN

I did a flight line check out at the Lakehurst Naval Air Station on a Raytheon Company radar warning receiver. No signal was detected about 45 degrees off the nose of the aircraft. I examined the photos that were taken for me. One photo was the front of the RU-21 aircraft with the nose section removed. The antenna dish of the AN/APR-215 was blocking the signal to one RWR antennas. That weather radar system antenna transverses 120 degrees in flight so there would be no blockage during flight.

The Raytheon Company was located in Santa Barbara, CA. Once Barbara accompanied me and we stayed in the Biltmore Hotel which was on the beach and had Government rates for rooms. Barbara was happy because she had access to a beach club that was across the street. She was raised in the ocean side town of Sea Bright NJ.

President Reagan's ranch was located atop the Santa Ynez Mountain which was above Santa Barbara. I drove up to the top looking for the ranch. There were driveways but no sign of the ranch.

President Reagan and I have the same birthday. So I sent a letter to the White House wishing him a happy birthday. My daughter Sharon was a volunteer in the White House and read the incoming mail. Most letters would be set aside for a standard reply and a "rubber stamp" signature.

I received a reply from the president dated 16 February 1984. The letter started with "Happy birthday" then his personnel comments. I believe he signed the letter because I previously received the standard reply in 1982 for an earlier letter and that rubber stamp signature was different. Also the Birthday letter was signed with a darker ink pen.

ADDENDA 35. HOW THE NUMBERS CAME TO BE

A college professor showed our class how the numbers came to be. Over the years I have embellished it and showed it to many others.

Thousands of years ago people tried to depict numbers but other people could not understand how to replicate those numbers.

Starting around 3,000 years ago Egyptians used images to depict numbers, e.g., A lotus plant image represented 1,000.

Roman Numerals originated in ancient Rome but that has been replaced with the easier to use Arabic numerals.

To replicate the Arabic numbers each number uses angles to represent each number. The number one was represented as a figure with one internal angle as noted:

This number two has two internal angles:

This number three has three internal angles:

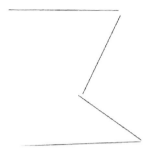

This number four has a triangle which has three angles plus one right angle:

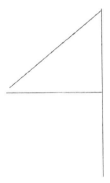

This number five has five right angles:

This number six has two right angles plus one rectangle.

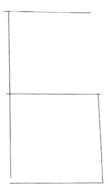

This number eight has two rectangles.

The number nine has one rectangle and five right angles.

A number is missing. For years that number was not an Arabic number. Then the number zero was added and the numbers became the Hindu Arabic numerals.

Another number is also missing. We usually do not use it but it is used in Europe and South America. A line is added to the seven so as not to confuse it with the one.

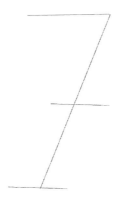

I gave a presentation about these number at our Computer Club meeting in our Adult Community. At the next meeting a member thanked me and stated that he gave the presentation to his grandchildren. I told him that it was not true and I forgot to tell that to the club.

ADDENDA 36. GERMANY SKIING

An Army captain was my liaison officer during a Travel Duty to Germany. On Saturday we drove to a Bavarian ski resort. The scenery in the Black Forest was spectacular. Half-way down a ski slope there was a kiosk that sold gluhwein. That mulled wine is warm red wine with added spices.

When our daughter Nancy was in the Eight Grade there was a ski trip to a beginner's hill. She enjoyed the skiing so I made a decision. I would stop playing golf and my family would ski.

On a ski trip we stayed in a Holiday Inn in Burlington Vermont. That night the temperature was near zero and in the morning our car would not start. A truck from a nearby gas station used battery cables to start my car and start other cars in the parking lot.

When we arrived at a ski lodge we were told that there was no available rooms at the lodge. We needed to stay in the Annex. Skiers in the Fort Monmouth Ski Club told me that they called the Annex the "shed." Our room was small and the only light bulb in the room would flicker. Barbara was afraid the shed would burn down during the night. The front door could not be locked and twice people tried to enter. This time I disconnected the car battery and carried it into our room.

At the Pico Mountain Ski Resort, There was a chair lift on a beginner's hill. Sharon would only ski on flat surfaces near the ski lodges. She did not want to use the rope tow at the other ski resorts. I asked the operator of that chair lift if he would stop the lift to let my daughter and I get off the chair. When Sharon and I were on the top of the hill I held Sharon between my legs and we transverse the slope. Soon Sharon was going on the chair lift and skiing down the hill without assistance.

Barbara did take a ski lesson and during our next trip she climbed halfway up the beginner's hill. She worried that she might hit the wooden rail fence rail at the bottom of the hill. I told her she could not travel that far. She did travel that far, hit the fence then never wanted to ski again.

We traveled to the Whiteface Ski Resort and stayed in the beautiful Lake Placid village. Jay and I traveled down the run with a bobsled driver. That bobsled run was the event site for the 1980 Olympics.

ADDENDA 37. MISSED DETECTIONS AND FALSE ALARMS

Electronic system designers need to determine an acceptable signal-to-noise (S/N) level. This includes Electronic Warfare jammers and detectors.

Our Universe generates White noise which is noise that is present at all RF frequencies. All receivers should have a Minimum Discernable Signal level high enough to detect desired signals above the noise but not too high to detect false signals. The system should be designed with a minimum RF bandwidth and a high gain antenna.

When our satellites map the surface of distant objects in space, the received data is stored in an on-onboard computer. Then the data is transmitted to very large parabolic antennas on Earth.

The pulse width of that transmitted signal is very wide so the receivers have very narrow bandwidths. This results in a higher S/N. That wide pulse width also results in a transmission that may require hours, days or weeks to complete.

ADDENDA 38. SHARING A TRAILER

The Electronic Warfare Laboratory building was a Top Secret building. There was a fence around that section of Fort Monmouth. EWL had a second fence around the building with a security guard in the lobby 24/7. It was a metal building without any windows. The building was a Faraday Shield. RF signals could neither enter nor exit the building.

EWL continued to hire more engineers and need more working space. They purchased a trailer and installed it inside the fence. I was one of the four contractors who would move into the trailer.

My assistant Bruce and I would share half the trailer with two other contractors. They were PhDs working on a nuclear radiation program.

I made a drawing of the trailer and divided into two equal sections. Then one of the PhDs pick one of two slips I held. They would have the

188 *John Edward Waters*

front of the trailer. They did not like having the section with the front door. I would not allow a Do Over.

I had a good relationship with both of them. I once mentioned to one of them that it is important in a marriage that both people like the same things. He said that it is also important that both people dislike the same things. (Interesting comment).

ADDENDA 39. PERSONAL NOTES

When I worked in the Counter Countermeasures (Anti Jamming) Branch, John a senior engineer asked me to review an article he prepared. I was pleased that he asked me to do it. I found a major error and showed it to him. He thanked me and made the correction.

An electronics magazine paid him $300 for the article. Conversely Sol who was in my car pool paid to have his articles published in technical journals. Sol was a division chief for electrical components. For him it was Publish or Perish.

Also at a nearby desk was Sam another senior engineer who was in charge of a recently published Engineering Handbook prepared by a contractor. He received a few calls from engineers who reported errors in the handbook. Sam told them he was a physicist and not an engineer. I would have reviewed the handbook if he asked me. I was a junior engineer but was once an Electronics Technician in the Navy and knew I could help him.

Sam previously was investigated by the Army McCarthy Hearings and was fired. He eventually was rehired.

A newspaper published my weekly articles on running, biking, swimming and triathlons. They only paid twenty dollars for each article. I did it as my contribution to those sports (and they also added my photo to each article). I asked Barbara to review each article. Her Fresh Eyes comments improved many articles.

If John had reviewed his article at a later date he would have noticed his error. I always review all my correspondence. Reviewing and revising

Addendum 1 through 39

189

my writings changed me from a poor writer to a good writer. When I asked our secretary to do a second revision I felt guilty for asking. Then all the engineers were given a personal computer and it was easy for me to do my own revisions.

When I traveled to Eglin Air Force Base and Fort Rucker, AL I flew on Southeast Airlines. That airline always had late departures. Southeast Airlines used the Eglin AFB runways. Once I had an early morning departure on Southeast Airlines and arrived at their small stand-alone terminal at Eglin. It was an originating flight and the plane was outside the terminal. Never-the less there was an hour delay from our scheduled departure.

Once I brought Barbara and our two children to Eglin and we stayed at nearby Fort Walton Beach. I booked a flight that was half empty the last time I flew. I planned to hold Nancy who was five years old and Barbara would hold Jay who was two years old. After we were airborne I would find two empty seats for Nancy and myself. That flight was full and the children were on our laps the entire flight.

My assistant engineer Ed always want to go to Eglin or Fort Rucker. He married a Southeast Airline flight attendant. My co-worker Bob married a flight attendant from a different airline.

The eighteen models of my Limited Production Countermeasures Set AN/ALQ-80 were ready for environmental tests at the Hallicrafters Company plant in Rolling Meadows, IL. The tests were estimated to be one month to complete. I had other priorities so a replacement was provided. I only knew that my replacement Bernie conducted a monthly Duplicate Bridge game that I planned to join in the future.

It required two months of testing to complete those tests. The contractor asked me to never let Bernie visit their plant again. I said OK. (Someday I may want them to do something for me).

I issued a Request For Proposal for a system that could jam a certain type of radar. Three proposals were received. Two were impractical for Army aircraft. The third would only generate range errors. Our current noise jammers generated range errors.

I did not award any contract. When visiting one of the contractors about another program I was asked about the canceled program. I did not say anything because I did not want to start a debate about the effectiveness of their system.

Our Division Chief did asked about the program and I did discuss it with him. At that time I had a major project and that program would take time away from the project.

I finished working in my office one Saturday afternoon and tried to close my safe. One of the four drawers in my safe would not close. I was the only person in the building except for the guard in the lobby. I did not want to get him involved

Cooper, the Fort Monmouth locksmith, once helped me and I had his office telephone number. No one answered when I called. Cooper always had his cap on and we used the same barber shop. I called our barber shop and told Bernie that I needed to call Cooper and asked if he knew where he lived. Bernie stated that he lived in Rumson and Cooper was in the shop and just left to go to a church in Long Branch. So calling his home would not help me.

I removed all the documents in that drawer then removed the drawer. Looking into the safe I saw a ball bearing on the bearing rail. I removed the bearing and was able to close the drawer.

I attended a lecture by Werner von Braun arranged by the Armed Service Communication and Electronics Association. He was the German American aerospace engineer who developed the rockets that launched the United States first space satellite. Under Operation Paperclip he could come to the United States. He said he liked the U.S. Lionel model trains and that helped him make his decision to come to our country.

During World War II, one of the technicians who worked for me landed in a glider a few day before soldiers landed on the Normandy beaches. I believe that Bob said he was in the 101st Airborne Division but he never told me what he did after they landed. I wanted to know but never asked him.

Bob had a problem with his back so I would ask someone else if I needed help lifting things. Once we were with a couple of other EWL

Addendum 1 through 39

191

personnel. As part of the conversation I stated that only one four leg animal has four knees. That animal is an elephant. Bob stated that a man has four knees. He has a left knee, a right knee, a hinee and a wenee. Since then I have told both statements many times. (Other four legged animals have two hocks).

My brother-in-law was in the 101st Airborne Division years ago. (I briefed their pilots during a meeting in Fort Campbell, KY)

Tom and I planned a ten day trip to visit Yellowstone National Park. At Newark airport we removed our bike pedals then placed the bikes in the boxes that the airline provided. We flew to Billings Montana then pedaled 500 miles in ten days in Yellowstone National Park and Montana and Wyoming

Prior to our trip I mailed a letter to the Chamber of Commerce in a town at the foot of Beartooth Mountain.

Their reply stated that in mid-September the Beartooth Mountain may have snow on it. I changed our arrival date to early September. (It snowed on the mountain the day after we traveled over it).

The mountain was more than 20,000 feet high. We had our old ten speed racing bikes and were not acclimated to high altitudes. So we walked up the last half mile. It was very cold on the top of the mountain and I needed to descend quickly. When Tom descended he was traveling forty miles an hour per his Cat Eye speedometer. I passed him.

During a breakfast in the park the server told us to beware of grizzly bears. Then he told us a scary story about a fisherman and the bear.

Later that day we saw a grizzly bear in the meadow walking towards us. We stopped and watched him approach. Then Tom wanted to take a picture with his camera. I recalled a saying, "If you do not want to be eaten by a bear then run faster than your brother." I told Tom that he could meet me at the top of the hill.

We pedaled across the Continental Divide. I stopped and poured water on the line. Some may flow into the Pacific Ocean and some may flow into the Mississippi River.

On our last day we shipped our panniers from a Greyhound bus terminal to be delivered to Billings. Then we biked one hundred miles on an interstate highway. Bikers can use interstate highways if there is no alternate route.

When our cat passed on it was a sad period. So I never wanted to have another cat. Many years later at my age that was unlikely to happen again. So on my birthday we purchased a cat from a Rescue Center.

Once we let in our house the neighbor's dog. Magic was used to being with cats and dogs at the center so it started as a peaceful meeting. Except when the dog began eating Magic's food then our cat jumped on him.

Recently Magic jumps in the car trunk when I remove the shopping bags. Now he wants to go in the back seat so I get a glass of wine and join him. That is the time he purrs the loudest.

My brother was the President of an American Legion Chapter. The chapter arranged for a newspaper to cover their Memorial Day ceremony. Bob asked my son Colonel Jay to be the speaker. Bob placed flowers at the memorial. He stated that is was one of the best days of his life.

I told my other brother that I was writing a book. He said he was thinking about writing a book. He worked with a doctor who was always drunk. I stated it could be a good story but he needs to write another hundred stories.

Addendum 1 through 39   193

My sister is a play writer and we have seen three of her plays. Gloria and her husband sold their home then rented an apartment in a large old house on the property of an ocean side beach club. The other renters in the house seem odd. (Gloria may be obtaining material for another play).

Our older daughter Nancy was a manger in a company in Manhattan. In Wall Street the company was considered a mystery company. An emir owned the company and other emirs and princes invested in the company.

The company researched and discovered companies that need money and who they consider to be mismanaged. Some of those companies were nationally known. Then they would take control of the company, fund it and replace the management. One of Nancy's duties was to hire people. She went to Harvard University and recruited their best students. She went to Eaton University and recruited their best students to work in their London Office.

When she asked to leave to spend more time with her children the company made her an offer. She could work at home and come to work one day a week. She accepted but was required to attend business meetings in London, Iceland, Dubai and their Headquarters in Bahrain.

When Nancy was a child a florist delivered flowers to our home. When I came home Nancy told me that flowers were delivered but we knew it was a mistake because you would never send Mommy flowers. The next week I asked a florist to deliver flowers and write on the card, "This is not a mistake."

Col Jay retired and volunteers as the Voices of Freedom Project leader with the Americans in Wartime National Museum.

Currently he is hiking the 800 mile Arizona National Scenic Trail for Warrior Expedition. You must be a veteran who has served in a combat zone to be part of the Warrior Expedition. The trail begins on the Mexican border and ends in Utah. Unlike the Appellation Trail where you see many other hikers, Jay sees only a few people and no hikers. His fellow vet dropped out after four days of hiking.

Our other daughter worked in Washington as a senior auditor for AmeriCorps. Sharon was also a volunteer working in the White House reading then filing incoming mail. Then Sharon received a scholarship

194   *John Edward Waters*

from New York University and graduated with a Masters in Journalism. She became the editor of the NJBIZ publication. Now at Montclair Stated University she is an adjunct professor and Program Manager for the Feliciano Center for Entrepreneurship. Sharon reviewed this manuscript.

My two brothers, my friends and I all served in the Army, Air Force or the Navy. My friend Bib was drafted during the Korean War. When Bib and the other recruits were sworn in at the In Processing Center, one recruit out of three recruits was assigned to serve in the Marines. Bib is our only Marine friend.

Wayne was an engineer working at Fort Monmouth. We are both magicians and are members of the International Brotherhood of Magicians, Ring 123. During the Christmas holiday period Wayne dresses as Santa and performs. Each year he performs at the Robert Wood Johnson Medical Center in New Brunswick NJ. The Center has a Christmas party for their children patients. Wayne's son was successfully treated for leukemia at the center. So that is Wayne's way to give back.

Wayne asked me to perform magic at each table during the parties. I watched as the children with major medical problems approached Santa... Some were pushing a stand with an IV bottle connected to the stand. That was an emotional time for me.

I entered a Chinese restaurant to purchase two egg rolls. A friendly customer started a conversation with me. He worked around the corner at an auto parts store. I told Hank that I know his supervisor and his wife Kutche cuts and styles my wife's hair. Barbara is bound to a wheel chair and cannot go to a beauty salon because they cannot wash her hair. So Barbara's live-aide washes her hair in the morning on her bed. After Kutche finishes working at a beauty salon she returns to our adult community then she comes to our home and cuts and styles Barbara hair.

Hank stated that he is going to school to study software programing. I think that he could have a successful career if he cuts his long hair and discards his T-shirt that has ZOMBIE printed on it.

There are statics that state that four year college graduates make a lot more money than non-graduates. But there are other factors to consider. Four year graduates are smarter than people who do not go to college.

Addendum 1 through 39

Using logic it is a given that students who pass a course in Latin are smarter that student who do not take Latin. But logic states that students who will take a course in Latin are smarter before taking the course.

It is known that STEAM graduates will start at a higher salary than Liberal Arts graduates.

Many Trade School students have rewarding careers. My electrician has other electricians working for him and he is a guest speaker each year at his trade school.

Scott, my plumber, has three plumbers working for him. One of those plumbers told me that he is going to a trade school at night. Scott once had a full time job at Costco. He also shoveled snow off my driveway. If our adult community has less than five inches of snow the community does not shovel driveways. So Scott would take that day off and shovel many driveways in our adult community.

ABOUT THE AUTHOR

John Edward Waters
Project Engineer
Toms River, NJ, US
Email: johnwaters85@gmail.com

After my Navy enlistment and college, I began my career as an electronic engineer in the Armed Service Electro-Standard Agency (ASESA), in Fort Monmouth, NJ. ASESA was moving to Dayton, Ohio so I visited Dayton. I told Barbara that the streets are wide and the people are friendly. She said there is the Pacific Ocean and the Atlantic Ocean but there is nothing in between.

Fortunately, I was selected for a position in the R & D Electronic Counter Countermeasures (anti-jamming) Branch. I designed and tested a circuit then published a report, ECCM circuit for Radar Set AN/PPS-5.

Then I was promoted as a Project Engineer in the R & D Electronic Warfare Laboratory reporting to the Program Manager Aircraft Survivability Equipment. I was a project engineer for Army aircraft electronic warfare systems for more than twenty years and also spent five years as a consulting engineer for other military systems.

INDEX

A

access, 179, 182
aerospace, 191
Afghanistan, 85, 175
Africa, 108
AH-1, 33, 34
Air Force, 25, 65, 99, 100, 121, 147, 148,
    151, 154, 157, 159, 170, 172, 174, 189,
    195
aircraft, 4, 17, 18, 19, 21, 29, 33, 35, 37, 43,
    47, 53, 55, 57, 73, 74, 77, 78, 91, 99,
    100, 101, 107, 109, 111, 114, 115, 116,
    119, 121, 123, 124, 129, 130, 134, 135,
    137, 143, 145, 146, 147, 148, 151, 153,
    154, 155, 156, 157, 158, 159,160, 163,
    164, 165, 170, 171, 174, 175, 176, 179,
    180, 181, 190, 199
airports, 129
Alaska, 91, 92
AN/APR-39, 29, 30, 31, 35, 37, 41, 43, 45,
    47, 77, 109, 111, 119, 121, 123, 137,
    141, 148, 154, 156, 157, 160, 164, 167,
    170, 175, 180
AN/PPS-5, 152, 199
AN/TPQ-36, 85

AN/TPQ-37, 87
AOC, 1, 143
arrest, 95
ASA, 147
ASESA, 166, 199
assessment, 152
assets, 164, 172, 176
audit, 162

B

bandwidth, 123, 149, 187
base, 57, 86, 91, 99, 100, 107, 109, 130,
    131, 133, 157, 159, 160, 171
beams, 121
beer, 155, 156, 175
Belgium, 115
boat, 39, 41, 138, 164
Border Patrol, 180

C

cabinets, 159
cables, 156, 186
Camp Evans, 1, 145
Canada, 25

Index

Carter, President Jimmy, 131
cartoon, 124
certification, 158, 166
chaff, 3, 4, 20, 53, 55, 160
Chamber of Commerce, 192
chicken, 157
children, 1, 89, 143, 159, 167, 190, 194, 195
CIA, 41, 175
circuit, 21, 27, 63, 74, 109, 119, 149, 150, 163, 175, 178, 180, 199
cleaning, 131
cockpit, 20, 34, 35, 73, 119, 120, 129, 135, 156, 164
COL, 117, 139, 173, 194
Cold War, 58, 68, 107, 146
collusion, 95
color, 24, 181
commercial, 21, 180
committee, 17, 110, 178
community, 66, 143, 144, 146, 196
computer, 104, 148, 149, 150, 159, 160, 186, 188, 189
conference, 105, 111, 150, 168, 170
conflict, 77, 115
Congress, 110
consulting, 87, 96, 104, 149, 161, 176, 199
Continental, 192
convention, 1, 89
copper, 179
cost, 87, 179
crimes, 95
crows, 2, 89, 90, 176, 196
Cuba, 159

D

D-Day, 5, 9, 11
de Gaulle, Charles, 107
deficiency, 113
Department of Defense, 167, 177, 197
detection, 171
devices, 41, 66, 81, 153, 160, 163, 177

diagram, 45, 161
doctors, 139
drawing, 5, 188

E

El Salvador, 41
emergency, 157, 171
employees, 30, 137, 162
endangered, 179
endangered species, 179
engineering, 49
England, 123
EPS, 162, 163, 179
equipment, 23, 24, 65, 73, 77, 87, 109, 111, 119, 123, 133, 135, 146, 147, 153, 158, 166, 199
Etcetera, 188
Europe, 186
evidence, 110
EWL, 66, 97, 149, 150, 152, 153, 155, 161, 162, 167, 171, 173, 176, 177, 179, 188, 191
exercise, 133, 134
explosion, 127

F

fascism, 117
FBI, 165
federal government, 110
field tests, 65, 180
figurine, 89, 90
filters, 123, 125
Firefinder, 85, 87, 88
flight, 23, 24, 49, 50, 65, 71, 77, 93, 99, 100, 121, 123, 130, 137, 147, 154, 155, 156, 157, 158, 159, 160, 167, 168, 169, 171, 172, 181, 190
flight attendant, 190
flights, 156
Fort Hood, 57

Index

FPB, 39, 41
France, 9, 107
fuel, 73, 77, 109, 111, 159
funding, 66, 68
funds, 30, 150, 172

G

gamma rays, 177
GAO, 110, 169
General Accounting Office, 110
Georgia, 103, 149
Germany, 4, 57, 58, 77, 107, 108, 109, 130, 157, 186
GPS, 41, 167, 168
graduate students, 107
Guardrail, 18, 148, 156
guidance, 47, 63, 170, 171
gun dish, 57, 58, 59, 119
gunpowder, 127, 152

H

Haiti, 168
Hawaii, 137, 138
highways, 50, 153, 192
history, 143
hobby, 124
Honduras, 129
horn, 163
House, 13, 14, 15, 41, 182, 195
hunting, 133

I

Iceland, 194
image, 47, 177, 182
in-board, 153, 154
indicator, 20, 33, 35, 37, 45, 63, 115, 130, 133, 135, 155, 156, 160, 164, 170
inspectors, 111, 169, 172
intelligence, 148

interference, 66, 154
Iran, 43, 44
Iraq, 81
IRCM, 174, 175, 176
Ireland, 124
Italy, 12, 81, 89, 96, 108, 117, 134, 165

J

Japan, 145

K

Korea, 47, 58, 73
Kuwait, 127

L

Lakehurst, 41, 50, 71, 93, 143, 157, 158, 181
lead, 4, 110, 168
LED, 124, 134, 135, 152, 163, 179, 180
leukemia, 195
lifetime, 100
light, 35, 63, 73, 74, 133, 135, 159, 170, 177, 187
link, 160
lithium, 180
livestock, 179
logistics, 17, 148
Lomell, 5, 7, 8
LTC, 88
luggage, 131, 169

M

management, 194
Marine Corps, 167
Maryland, 13, 41, 51, 88, 104, 158
MDS, 149
measurements, 162, 163
media, 49, 83

Index

medical, 195
Middle East, 95
military, 18, 25, 55, 58, 65, 77, 91, 93, 95, 99, 104, 105, 108, 110, 119, 121, 124, 129, 131, 133, 137, 143, 146, 150, 162, 165, 172, 175, 199
Minneapolis, 169
missile, 20, 21, 35, 47, 58, 59, 61, 63, 65, 67, 134, 135, 168, 170, 171, 174, 175, 197
missions, 4, 43, 51, 73, 100, 103, 105, 121, 133, 151, 157, 171, 174
Mississippi River, 192
Missouri, 107, 119
models, 30, 57, 74, 109, 146, 147, 156, 190
Morocco, 55
Moscow, 83

N

NAS, 71
National Security Agency, 41
Nellis AFB, 155
neutral, 144
niche market, 161
night vision, 133, 135
North America, 155
North Korea, 47
Norway, 123
numbers, 182, 185

O

officials, 110
OH-6, 49, 164
operations, 148, 168

P

Pacific, 151, 192, 199
Panama, 131, 132
parallel, 50, 160

peace, 77
peacekeepers, 168
Pearl Harbor, 137, 138
pencil, 51
Pentagon, 25, 110
Philadelphia, 156
Philippines, 99
photographers, 158
plaque, 13, 14, 51, 141
playing, 49, 186
PM, 119, 121, 123, 131, 137, 147, 148, 174
population, 158, 177
prisoners, 43
procurement, 17, 30
profit, 110
project, 29, 41, 66, 68, 71, 77, 79, 91, 97, 105, 115, 124, 146, 149, 153, 157, 159, 161, 167, 173, 179, 190, 199
propaganda, 49
publishing, 49
PVC, 180, 181
pylon, 73, 75, 146, 174

Q

QRC, 172, 173

R

racing, 192
radar, 4, 19, 20, 21, 23, 24, 27, 29, 31, 33, 34, 35, 37, 39, 41, 43, 45, 47, 55, 57, 58, 63, 66, 73, 74, 77, 93, 95, 100, 103, 113, 114, 115, 116, 119, 121, 123, 124, 129, 130, 133, 137, 145, 146, 147, 151, 152, 153, 154, 155, 156, 157, 158, 159, 160, 161, 162, 166, 170, 171, 172, 174, 178, 179, 180, 181, 190
radiation, 166, 170, 176, 177, 178, 188
radiation, 176, 177
radio, 145
rangers, 5, 6

Index

reading, 83, 96, 195
reliability, 99, 177
reporters, 50
retirement, 51, 139, 141, 142, 143
Ronald Reagan, 43, 181
Rucker, 113, 114, 123, 189, 190
Russia, 92

S

SA-2, 47, 58, 63, 170, 171
SA-7, 61, 174, 175
scholarship, 93, 195
school, 85, 88, 93, 196
SCIF, 83
Secret Service, 14, 15
Secretary of Defense, 146
security, 99, 129, 150, 188
sensor, 148, 177, 179, 180
sensors, 176, 179, 180
shortage, 111
SIGINT, 148, 149, 156
signals, 20, 21, 23, 29, 33, 35, 37, 47, 48, 63, 65, 79, 104, 109, 113, 121, 145, 146, 147, 148, 149, 151, 152, 155, 162, 164, 166, 167, 168, 170, 180, 181, 187, 188
signs, 18, 130, 133
skeleton, 139
slingshot, 93, 94
smoking, 171, 178
software, 149, 196
sortie, 170
Soviet Union, 146, 147
space, 65, 159, 188, 191, 197
starvation, 49
state, 17, 30, 93, 105, 156, 196
stock price, 160, 176
submarines, 155, 171
supervisor, 14, 30, 39, 66, 95, 104, 105, 127, 153, 163, 168, 179, 196
surplus, 167
surveillance, 19, 73, 111

T

tactics, 151, 170, 172
target, 4, 63, 87, 130, 156, 174
technical assistance, 129
technician, 65, 109, 130, 152, 161
telephone, 41, 95, 96, 129, 175, 191
temperature, 91, 164, 181, 186
test items, 24
testing, 29, 71, 74, 190
Thailand, 99, 100
tracks, 50, 79
trailer, 163, 170, 188
training, 91, 103, 133, 160
transactions, 107
transistor, 27, 110
transmission, 66, 154, 168, 188
transport, 73, 75
transportation, 129
triangulation, 116
truck, 65, 67, 153, 177, 186

U

U-2, 17
undercover, 95
uniform, 144, 152, 171
United Kingdom, 123
United States, 43, 83, 91, 191
Unmanned Aerial Vehicles, 104, 176
urban, 153
USA, 138

V

V1, 45, 47, 48
V2, 47, 48, 175
vacuum, 152
vehicles, 162, 168, 180, 181
velocity, 79

204 *Index*

Vietnam, 49, 61, 63, 74, 99, 101, 119, 170, 171, 172, 174, 179
vision, 133, 135
volunteer work, 14, 195
VT fuse, 127, 147

W

walking, 133, 192
war, 4, 57, 105, 108
warning systems, 151, 155
Washington, 1, 89, 97, 103, 104, 195

waste, 167
water, 161, 165, 172, 192
water heater, 161
waters, 9, 50, 139, 141, 161, 165, 172, 179, 192, 199
weapons, 130
White House, 13, 14, 15, 182, 195
wires, 119, 178
wood, 158
World War I, 3, 5, 11, 57, 58, 91, 117, 144, 177, 191

Related Nova Publications

LIQUID FUELS AS JET FUELS AND PROPELLANTS: A REVIEW OF THEIR PRODUCTIONS AND APPLICATIONS

AUTHOR: Mohammad Hossein Keshavarz, Ph.D.

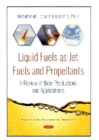

SERIES: Energy Science, Engineering and Technology

BOOK DESCRIPTION: This book reviews some efforts that have been done to introduce new candidates for replacing conventional hydrazine fuels because they are acutely toxic and suspected carcinogens, costly safety precautions and handling procedures are required.

SOFTCOVER ISBN: 978-1-53614-311-9
RETAIL PRICE: $82

ADVANCES IN MATERIALS SCIENCE RESEARCH. VOLUME 33

EDITOR: Maryann C. Wythers

SERIES: Advances in Materials Science Research

BOOK DESCRIPTION: In this compilation, bismuth-based ceramics having chemical formula $0.96|\{Bi0.5(Na0.84K0.16)0.5\}1-xLix(Ti1-yNby)O3]-0.04SrTiO3$ (BNKLiTN–ST) with $x=y = 0$–0.030 were synthesized by a conventional solid-state reaction method, and a stack-type multilayer actuator was fabricated by using optimal composition.

HARDCOVER ISBN: 978-1-53614-313-3
RETAIL PRICE: $250

To see a complete list of Nova publications, please visit our website at www.novapublishers.com

Related Nova Publications

UNITED STATES NAVY: BACKGROUND AND ISSUES FOR CONGRESS

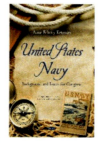

AUTHOR: Aune Felicity Krizman

SERIES: Congressional Policies, Practices and Procedures

BOOK DESCRIPTION: This book provides background information and potential oversight issues for Congress on the Columbia-class program, the Gerald R. Ford (CVN-78) class aircraft carrier program, the Navy's FFG(X) program, on the Navy's Littoral Combat Ship (LCS) program and on three new ship-based weapons the Navy is developing that could improve the ability of Navy surface ships to defend themselves against missiles.

HARDCOVER ISBN: 978-1-53614-767-4
RETAIL PRICE: $230

HUMAN COLLABORATION IN HOMELAND SECURITY (DVD INCLUDED)

EDITORS: Robert Irving Desourdis and Kuan Hengameh Collins

SERIES: Homeland Security and Safety

BOOK DESCRIPTION: Part I of our book explains to both the general reader and homeland security experts alike, what individual and organizational factors are needed to establish a collaborative environment.

HARDCOVER ISBN: 978-1-53611-935-0
RETAIL PRICE: $310

To see a complete list of Nova publications, please visit our website at www.novapublishers.com